OPTICAL MICROSCANNERS AND MICROSPECTROMETERS USING THERMAL BIMORPH ACTUATORS

MICROSYSTEMS

Volume 8

OTHER BOOKS IN THE SERIES:

Methodology for the Modelling and Simulation of Microsystems
Bartlomiej F. Romanowicz
Hardbound, ISBN 0-7923-8306-0, October 1998
Microcantilevers for Atomic Force Microscope Data Storage
Benjamin W. Chui
Hardbound, ISBN 0-7923-8358, October 1998
Bringing Scanning Probe Microscopy Up to Speed
Stephen C. Minne, Scott R. Manalis, Calvin F. Quate
Hardbound, ISBN 0-7923-8466-0, February 1999
Micromachined Ultrasound-Based Proximity Sensors
Mark R. Hornung, Oliver Brand
Hardbound, ISBN 0-7923-8508-X, April 1999
Microfabrication in Tissue Engineering and Bioartificial Organs
Sangeeta Bhatia
Hardbound, ISBN 0-7923-8566-7, August 1999
Microscale Heat Conduction in Integrated Circuits and Their Constituent Films
Y. Sungtaek Ju, Kenneth E. Goodson
Hardbound, ISBN 0-7923-8591-8, August 1999
Scanning Probe Lithography
Hyongsok T. Soh, Kathryn Wilder Guarini, Calvin F. Quate
Hardbound, ISBN 0-7923-7361-8, June 2001

Optical Microscanners and Microspectrometers using Thermal Bimorph Actuators

Gerhard Lammel
Sandra Schweizer
and
Philippe Renaud

EPFL, Swiss Federal Institute of Technology Lausanne

KLUWER ACADEMIC PUBLISHERS
BOSTON / DORDRECHT / LONDON

A C.I.P. Catalogue record for this book is available from the Library of Congress.

ISBN 0-7923-7655-2

Published by Kluwer Academic Publishers,
P.O. Box 17, 3300 AA Dordrecht, The Netherlands.

Sold and distributed in North, Central and South America
by Kluwer Academic Publishers,
101 Philip Drive, Norwell, MA 02061, U.S.A.

In all other countries, sold and distributed
by Kluwer Academic Publishers,
P.O. Box 322, 3300 AH Dordrecht, The Netherlands.

ABOUT THE COVER

Left: Scanning electron microscope (SEM) picture of a wafer with an array of 2D micromirrors with "L"-shaped actuator beams, developed by Sandra Schweizer. The dry-release process leads to a high production yield.

Right (series of four pictures): Tunable optical filter of porous silicon in motion, developed by Gerhard Lammel. Under a voltage of 0 to 2.1 volt the plate turns by 90° and changes the filter wavelength.

Printed on acid-free paper

Printed in the Netherlands.

TABLE OF CONTENTS

PREFACE

This book is based on research results of Sandra Schweizer and Gerhard Lammel during their PhD thesis' at the Swiss Federal Institute of Technology Lausanne in the group of Prof. Philippe Renaud.

1 INTRODUCTION

1.1 MOTIVATION

In this book, we explore the application of MEMS technologies in micro-optical devices that use integrated micro-actuators for scanning mirrors and optical interference plates. These devices, as part of the so-called MOEMS (Micro Opto Electro Mechanical Systems) family, are primarily intended to be used as key components for linear and two-dimensional light beam scanning as well as for tunable optical filters.

We first present a thorough study of thin film thermomechanical actuators that are based on multimorph beams heated by resistive heating. Thermomechanical actuators have been chosen because they are very simple to integrate in thin film technologies and also because this transduction mechanism scales quite favorably with the miniaturization of the devices. Analytical models and numerical simulations are used to predict the behavior of the actuators and to extract design rules for the best performance for given sets of layers.

The next chapters are dedicated to beam steering applications. We present a variety of designs and technologies for the realization of scanning micro-mirrors. In particular, a very innovative concept for 2D scanning is introduced. It is made of only one actuator with a special shape that is able to generate simultaneously a low frequency movement that corresponds to frame scans and an orthogonal high frequency resonant mode that defines the line scans, see Fig. 1.1. This beam scanner can be used for light reflection imaging (i. e. barcode capture) or for micro-display applications (e. g. retina projection).

Then, the microscanner concept is complemented with optical interference filters that are made of porous silicon multilayers. Such technology opens the way for complex devices capable of tunable filtering of light (visible and IR). In these devices the thermomechanical actuator, partly composed of a porous layer, is manufactured within the same process of modulated electrochemical porosification for the multilayer bandgap mirror. This leads to a simple process integration. As an example, a single detector infrared spectrometer for gas detection is shown, Fig. 1.2.

In summary, this book gives a complete approach of design and technologies for micro-scanning devices that use thermomechanical actuation. The modulated porous silicon technology is thoroughly presented, including optimization and control of process parameters. The interest of these concepts and technologies is demonstrated by the case studies. But the same "toolbox" can certainly be applied for many other micro-optical device. And this is the objective of the book – to provide the reader with all necessary information to do so.

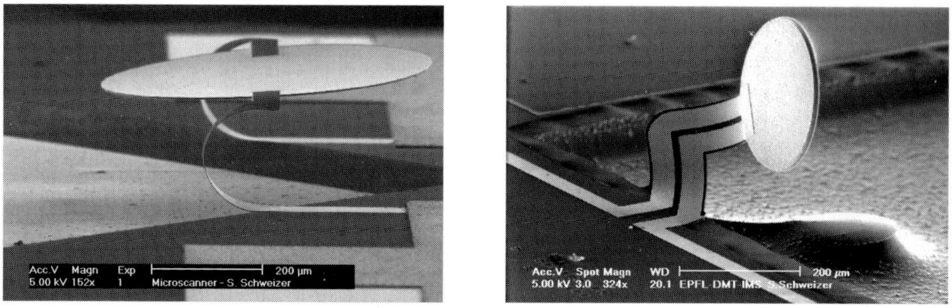

Fig. 1.1 *Thermally actuated microscanner with large angle and low power consumption. Left: version for 1-dimensional scanning, Right: version for 2-dimensional scanning with "L"-shaped beam. In the rest position the mirror is turned out of the chip plane by residual stress, S. Schweizer, EPFL, Switzerland.*

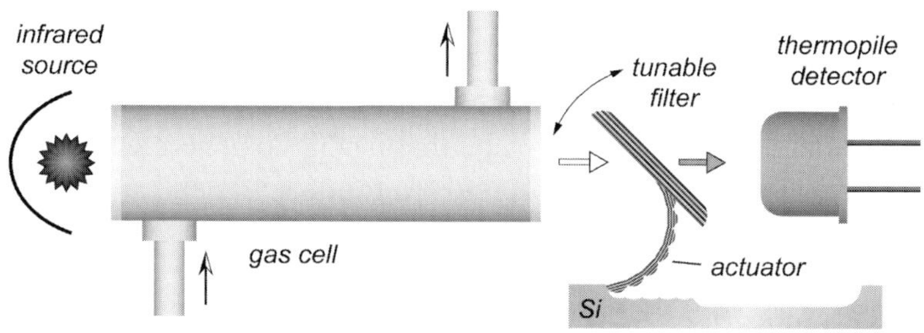

Fig. 1.2 *Operating principle of the MOEMS infrared gas spectrometer. It measures the infrared absorption of a gas at different wavelengths, which can be selected by a tunable interference filter, G. Lammel, EPFL, Switzerland.*

1.2 MEMS

In the past two decades, the technologies and applications of micro-electro-mechanical systems had a tremendous growth. Micro-electro-mechanical systems, or MEMS, are made of micromachined components and combine electrical and non-electrical functions in very small space. The core technology used in MEMS is derived from the planar lithographic technologies of integrated circuits with the addition of processes that permit the fabrication of mechanical, optical, fluidic or other components.

Among the possible advantages of MEMS compared to their classical counterparts, we can mention: cost and size reduction, increased reliability, less material used, batch processing for mass production, low power, etc. Of course, MEMS cannot, and will not, replace all existing devices. But many types of micro-sensors and some micro-actuator systems are partly or fully based on MEMS technology. This is a rich field of research and development in which some applications have already reached the commercial level and some open up very promising markets.

One of the best known example of MEMS is the integrated accelerometer. This system combines in a single miniaturized module a micro-mechanical spring-suspended proof mass, the sensing transducer (usually capacitive or piezoresistive) and a complex signal processing unit that amplifies, linearizes and digitizes the sensor output. It can also include self-test and self calibration functions.

1.3 MOEMS

The micro-opto-electro-mechanical systems (MOEMS) are a sub-division of MEMS that is commonly applied to microsystems dealing with optical functions.

Scanning, switching and deflecting an optical beam is required in many applications. A real explosion of new creative solutions was reported over the last five years, due to a dramatic demand in telecommunications with the development of WDM (Wavelength-Division-Multiplexing) systems. The field of micro displays also developed new components to satisfy the need of fast mobile communication devices with high information content.

A non-exhaustive list is given below, concentrated on MOEMS applications. Some of them found the way to industry and were commercialized; others, developed later on, but also very promising, have not reached maturity yet.

1.3.1 Display technologies

Micromachining technology has already shown that it has the potential to enable a new generation of projection video displays with the commercialization of the Texas Instruments' Digital Micromirror Display (DMD™). Commercialized and emerging micromachined projection-video-display technologies can be separated into three categories: the two-dimensional discrete-pixel arrays, the raster scanning systems, and the combined approach: the Grating Light Valve (GLV™) technology.

2-dimensional discrete-pixel arrays

A two-dimensional panel associates one physical pixel on the modulator array with each addressable spot on the screen (Fig. 1.4 a). One reflective element corresponds to one pixel. Depending on the type of modulator, each pixel operates by altering between transmissivity or reflectivity. Examples of display technologies using two-dimensional panels include liquid-crystal displays (LCD), digital micro-mirrors (DMD), and liquid-crystal-on-silicon (LCOS) [1].

This technology dates back to the 1970's. Petersen *et al.* [2] demonstrated reflecting light beams by small cantilevers driven by electrostatic force in 1977. A 16-element linear light modulator array composed of the cantilevers was used in a small display system. A line of He-Ne laser light was reflected by the cantilever, apertured, focused and scanned by a galvanometer.

Texas Instruments has introduced the Digital Micromirror Display (DMD™). The microreflector in the DMD consists of a small (about 20 x 20 μm^2) deflectable mirror that modulates the light by deflecting the beam in and out of the viewing aperture (Fig.1.3 a). A field of almost one million micromirrors is needed to produce a full SVGA (Super Video Graphics Array) quality display. At the size of about 20 x 20 μm^2 each, the display module will occupy a chip with an overall size of several square centimeters [3] (Fig. 1.3 b).

a) b)

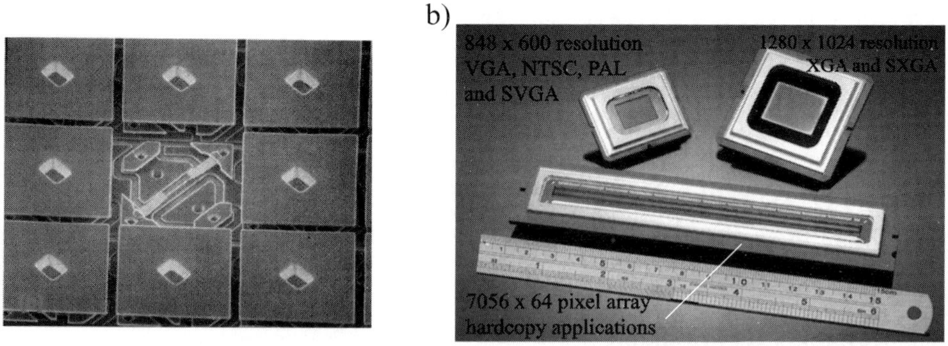

Fig. 1.3 a) 3 x 3 array of Texas Instruments DMD mirrors, where one mirror has been removed to reveal the underlying mechanical structure. b) Texas Instruments DMD chip configurations. Courtesy of L. Hornbeck [3].

Several academic groups have also developed micromirror arrays where each element is a mirror segment that serves as a pixel in a larger display [4-6]. For such systems it has been demonstrated that the fabrication yield of simple micromechanical actuators can approach 100%. Further, it has been established that large-scale integration of electronics with MEMS array structures is achievable. This integration has been performed by building the MEMS structures on top of a planarized CMOS electronics array, which is also the case for Texas Instruments' DMDs.

The benefits of the dicrete-pixel architecture include sharp transitions between pixels, a display resolution limited only by the panel geometry, and low-bandwidth video channels because each pixel is updated only once per frame. The principal disadvantage of the discrete-pixel array is the large number of pixels and video channels that must be fabricated within tight uniformity tolerances to achieve an acceptable image. Other disadvantages include a "screen door" (or fixed-pattern noise) effect around pixel boundaries and the complex electronics required to reformat an image between various aspect ratios.

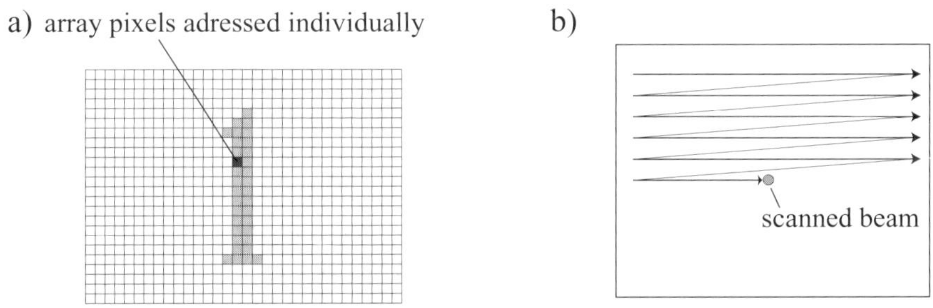

Fig. 1.4 Display techniques after Corrigan et al. [1]: a) Two-dimensional panel architecture addresses each pixel separately. One pixel on a modulator array is associated with each addressable spot on the screen. b) Scanned-beam architecture uses a single pixel moved in two dimensions to create an image.

Raster scanning systems

The scanned-beam architecture creates a single bright pixel that is scanned in two dimensions – in the same manner as a cathode ray tube (CRT) is scanned – to create an image (Fig. 1.4 b). These systems use two mirrors scanning in orthogonal directions or one mirror scanning in two dimensions. Raster scanning systems need a modulated light source to generate a 2D image. The inherent advantages of this architecture include the absence of pixel boundaries, the ability to synchronize the system to a wide ratio of video input signals, and the ability to easily reformat image aspect ratios. One intrinsic disadvantage is the necessity of very high bandwidth that must pass through a single video drive channel.

Sophisticated surface-micromachined mirrors with monolithically integrated electrostatic combdrive actuators have been reported for raster-scanning applications at the University of Berkeley (Berkeley Sensor & Actuator Center, BSAC) [7]. This has been extended to a full-motion video display application [8] implementing two mirrors (Fig. 1.5). The latest beam deflector of the Fraunhofer Institute of Dresden consists of one mirror and oscillates in two dimensions, electrostatically driven up to 20 kHz in one direction [9] (Fig. 1.6).

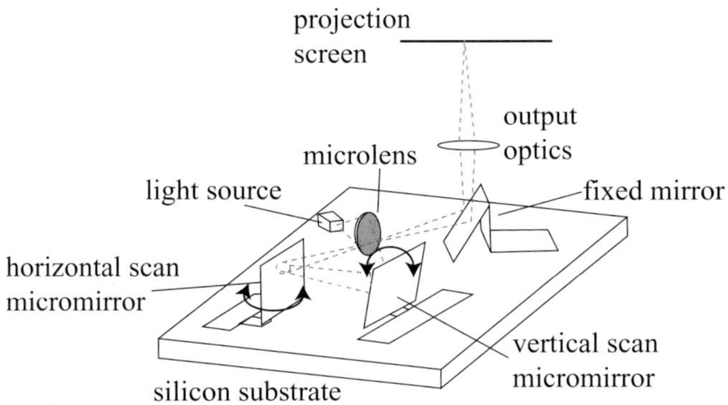

Fig. 1.5 Schematic of BSAC's integrated scanning-display module on a silicon substrate. The output from a laser diode or an LED is collimated and directed to the two scanning mirrors with orthogonal directions, and a final fixed mirror deflects the beam perpendicularly to the chip surface. Adapted from Kiang et al. [7].

A combined approach: the Grating Light Valve (GLV™) technology

Silicon Light Machine Inc. markets displays that use Grating Light Valve technology. The GLV™ display contains an external bulk-optical scanning mirror (a low-cost galvanometric scanner) and a one-dimensional array (Fig. 1.7 a) of micromachined deformable diffraction gratings that modulate the pixel intensity (see section 1.4.1, Fig. 1.17). The scanned linear architecture uses a linear array of physical pixels oriented along a vertical or horizontal column of image data (Fig. 1.7 b). Once per image frame, this linear array is optically scanned across the screen to produce a complete image.

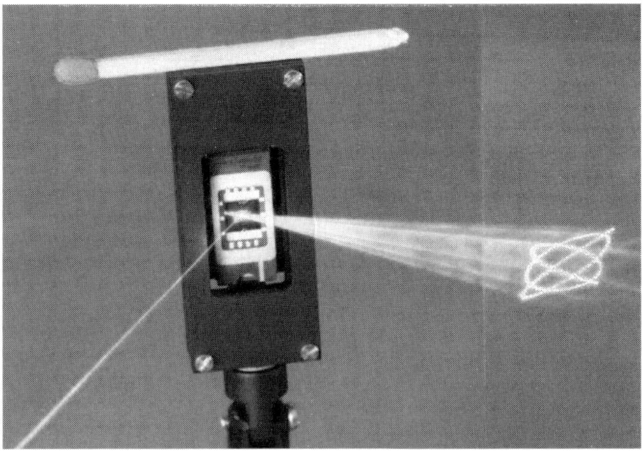

Fig. 1.6 2D scanner in operation, courtesy of H. Schenk et al., Fraunhofer Institute Dresden, Germany.

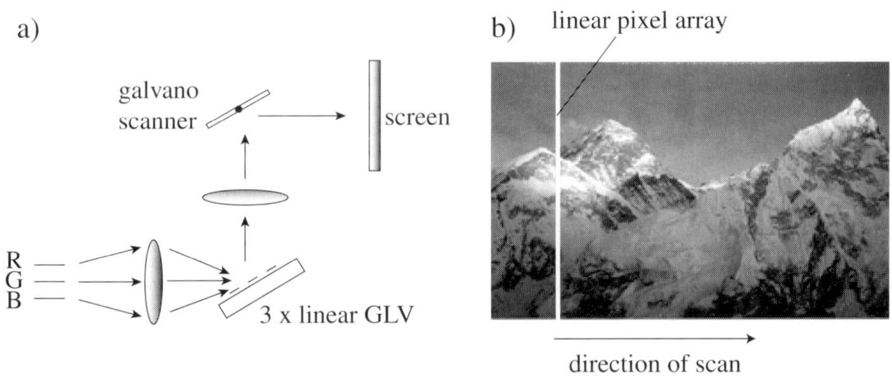

Fig. 1.7 a) Schematic of the GLV technology: Silicon Light Machine has demonstrated a laser-driven projection system where three separate GLVs modulate three light beams. b) In a scanned linear display architecture, a linear array of pixels is optically scanned across the screen to produce a complete image. Adapted from Corrigan et al. [1].

For SVGA quality displays, the linear diffraction grating array should contain up to 1000 elements and measure several centimeters in overall length. The bulk optical scanning mirror has similar dimensions.

Unlike the raster-scanning architecture, the linear-scanning system scans at the frame rate, rather than the line rate. The lower scan rate translates directly into less expensive and simpler optical scanning. The number of physical pixels in a scanned linear array is three orders of magnitude less than that of a two-dimensional discrete-pixel array, which translates directly into higher manufacturing yields and lower modulator costs. The scanned linear architecture supports extremely high resolutions because the size of the array scales in a linear fashion rather than geometrically.

Discrete pixel versus raster-scanning: micromirror features

There are fundamental differences in the requirements for micromirrors used in the DMD (discrete-pixel) and those used for raster-scanning display (the GLV is actually an amalgamation of the two approaches). The main difference is in the size of the micromirrors. The discrete-pixel mirrors act as a point source of on/off light scatters and thus can be made quite small (16 μm in the case of Texas Instruments' DMD), while the dimension of the raster-scanning mirror will determine the overall resolution of the displayed image according to the diffraction limit.

To produce 1000 resolvable spots over a 30° field, the diffraction limit dictates a minimum mirror size in the order of 0.5 mm. On the other hand, the need to scan over 1000 lines of an entire image frame at the video rate of 60 frames/s requires a scanning frequency of 60 kHz for the fast mirror (the orthogonal slow-mirror scans at video frame rate). At the same time, to maintain diffraction-limited performance, the optical quality of the mirror must be maintained to a minimum dynamic flatness of at least $\lambda/20$ over the entire mirror surface, while the mirror is scanning at high speed.

The "discrete-pixel" and the "raster-scanning" approaches are radically different. For the former, the demands are comparatively relaxed on the performance of each individual mirror, but requires a massive number of them to work simultaneously with exceptionally high yield. The latter demands high performance of the micromirror, but needs two or even only one of them to perform the display function.

It is the raster-scanning display, however, that offers the possibility of truly integrating the entire display engine on a chip, containing just one or two micromirrors with integrated light-sources and microlenses.

Other display applications

Besides improving the resolution and the brightness in classical applications such as television (HDTV, High Density TeleVision), VGA (Video Graphics Array) displays or projection systems (home theater and electronic cinema), for which the applications of micro light deflectors have already been demonstrated with Texas Instruments' DMD™ (Digital Micromirror Display), many creative new applications seem to be obtainable thanks to MEMS technology. A particular adaptation involves scanning the image directly on the retina of the viewing person, a technique known as *Virtual Retinal Display* (VRD) [10], which has been explored at the Human Interface Technology Laboratory of the University in Washington and currently attempted to be commercialized by Microvision Inc. This approach is promoted as one that has "see-through" capability, i.e., the displayed image can be viewed with a bright external scene superimposed on the image – a significant advantage in cockpit displays and industrial / medical applications. It also has the potential of being wearable since only one or two micro-mirrors are required to produce the entire image by raster scanning, the display hardware is truly miniaturized and fits easily onto an eyeglass frame [11, 12].

Micro-mirrors can be used as a deflector for angle-multiplexed *holographic storage* [13]. In general, volume holographic data storage is achieved by using two laser beams (one reference beam and one signal beam), which interfere with each other and form a 3-D interference pattern (hologram) in a light sensitive crystal that "memorizes" it. The reference beam is deflected by the MEMS mirror (section 1.4.3, Fig. 1.18 a), which offers low cost, low mass and large scanning angle.

1.3.2 Printers and industrial machining

To replace polygonal scanners and provide low cost at low operation power devices, micromirrors can find application in laser printers industry.

In high resolution optical applications, maskless lithography or fast submicron laser direct writing can be provided with deformable micromirror devices as phase light modulating valves [14]. In the same way, deformable micromirrors can be applied in laser controlled cutting torches for industrial high resolution machining [15].

1.3.3 Imaging scanners

Barcode scanners are widely used for automatic object identification in many industries. Rockwell Automation demonstrated a scan engine based on a bimorph thermal actuator micromirror integrated with a commercial barcode reader / decoder and successfully proven to read barcodes [16] (section 1.4.4, Fig. 1.20). Berkeley's silicon surface-micromachined scanning micromirrors with electrostatic comb drives (Fig. 1.8 a) [17], piezoelectric driven silicon micromirror integrated with a photodetector [18] or LETI's translating micro-lenses [19] (Fig. 1.8 b) have many advantages compared to bulk, galvanometric-type or rotating-polygon-type scanning mirrors. They have a fast scan rate, a large scan angle, low mass, and low operation power. More importantly, the integration of the scanner optics, driving motor, and processing electronics onto a single chip can be carried out as part of the fabrication process, greatly reducing assembly cost and improving system reliability [20].

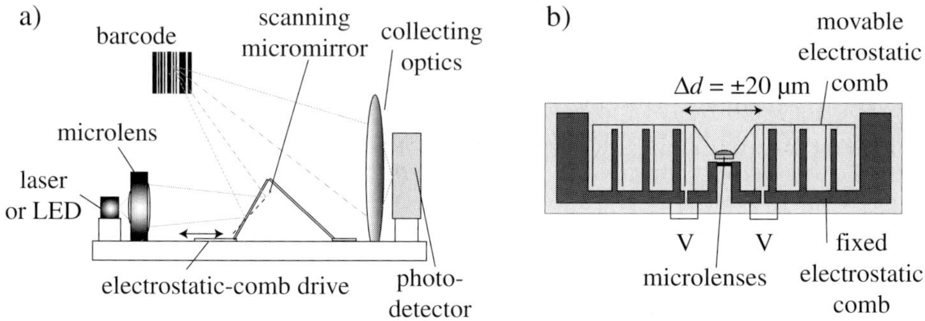

Fig. 1.8 a) Schematic of BSAC's integrated barcode-scanner module on a silicon substrate. After Kiang et al. [17]. b) Principle of LETI's scanning device. The laser beam deflection is achieved by an afocal system of microlenses. The electrostatic actuator is designed to scan the laser beam by translating movable microlenses in the orthogonal direction to the beam propagation. Chip size: 1.4 mm x 5 mm. Adapted from Petroz et al. [19].

An alternative system for barcode reading with potential for miniaturization uses a moving fibre or waveguide. Light emerging from the fibre is imaged to a spot by a lens, and a small displacement of the fibre tip is magnified to produce a scan line in the image plane. This displacement is generated by exciting the fundamental resonance of a fibre cantilever using, for example, piezoelectric actuation [21].

In many other sensing applications such as detectors for molecular beams [22], data storage, optical recognition sensors or laser radar systems, micro-deflectors and micro-scanning systems promise many advantages compared to current systems.

1.3.4 Telecommunication

One emerging application of MOEMS with favorable economic perspectives are optical switches for telecommunication. Although micro mechanics is too slow to route data packages in optical networks, optical switches are needed to reconfigure it. Networks have to adapt to the data traffic situation and allow to circumvent broken nodes.

A typical application is a *n* x *n* crossbar switch that connects any of the *n* input channels to any output. The requirements are low insertion loss (< 2 dB) and low channel crosstalk (< -60 dB).

Early devices used torsion mirrors supported by thin polysilicon beams for a 2 x 2 switching matrix [23]. The chip has connections for optical fibres, terminated by ball lenses [24]. A more elegant 2 x 2 optical switch design shifts the mirror into and out of the beam [25]. The advantage is that the fibres can be coupled so closely together that no focusing lens is required, see Fig 1.9.

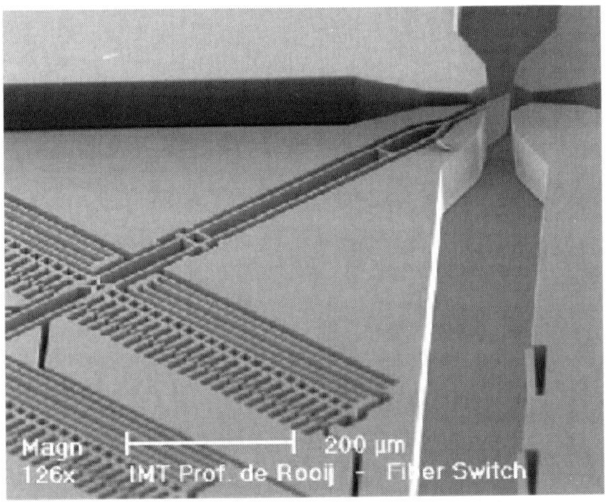

Fig. 1.9 *2 x 2 optical fiber switch with shifting mirror (top right), actuated by an electrostatic combdrive (bottom left). Courtesy of C. Marxer, IMT Neuchâtel, Sercalo microtechnology inc., Switzerland.*

The variety of design geometries for optical switches includes mirrors mounted on a rotating micro-wheel driven by a wobble motor [26].

Aksyuk *et al.* from Bell Labs Lucent Technologies presented 2-axis tilting micromirrors for free-space optical crossconnects of up to 256 x 256 fibres [27]. The round mirrors with 500 µm diameter are suspended in gimbal mounts via torsion springs and bearings and actuated electrostatically. The structure is lifted out of plane with fixed bimorph beams, see Fig. 1.10.

Neumeier *et al.* reported on an innovative approach that uses total reflection instead of mirrors to deflect light with low loss. A prism shape cavity in a

Fig. 1.10 Surface micromachined two-axis tilting mirror (500 μm diameter) for free space optical crossconnects. Courtesy of Bell Labs Lucent Technologies.

polymer can be filled with an index matching fluid like glycerine to become transparent or emptied to switch the beam by total reflection [28].

Agilent Technologies combined this concept with thermal inkjet printer technology to create bubbles in a channel, see Fig. 1.11. The silica planar waveguide is transparent when the index matching fluid is present in a trenched channel that crosses the lightpath. A gas bubble can redirect the beam by total reflection. The switching time is < 10 ms, the channel crosstalk < -50 dB, the average insertion loss 5 dB. Switching platforms of 32 x 32 fibres can be cascaded [29].

1.3.5 Adaptive or corrective optics

In adaptive or corrective optics, it is important that the deformable micro-mirror is both continuously and precisely adjustable. Micromachined adaptive mirrors combine low cost and simplicity of control with the possibility of mass production. These micro-deflectors find applications in telescopes [15] (section 1.4.1), in optical aberration control for optical disk storage [30] and micro-focusing devices for barcode readers for increased resolution and a wide sensing range [31].

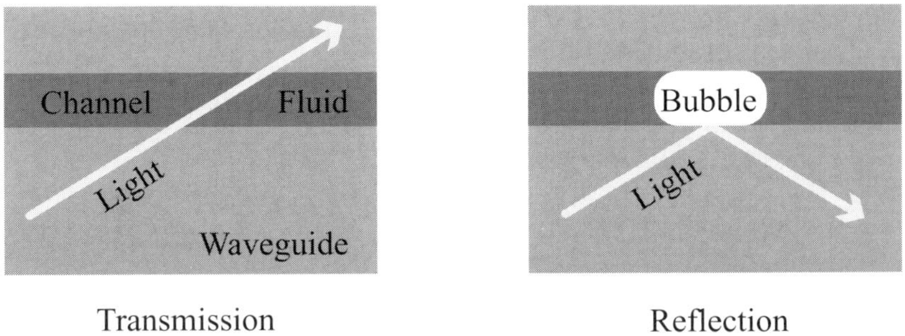

Fig. 1.11 Fluidic light switch: The lightbeam is transmitted in a waveguide when the crossing channel is filled with a fluid (left). A thermal inkjet element creates a bubble, leading to total internal reflection (right).

1.3.6 Spectroscopy applications

Tunable optical filters

Tunable optical filters find numerous applications in microsystems. In optical telecommunication, the data rate of a single fiber can be increased to a multiple by Wavelength Division Multiplexing (WDM). Up to 128 optical channels are transmitted simultaneously, separated by only 1 nm or less in wavelength. Data rates of more than 1 Terabit per second have been achieved with this technology. To add or drop a channel, wavelength selective filters are needed.

Strassner *et al.* reported on a Fabry-Pérot filter of two dielectric mirrors separated by an air gap, see Fig. 1.12. The upper mirror can be pulled down by an electrostatic actuator to adjust the size of the gap. This shifts the transmission wavelength by 45 nm, centered in the fiber optics telecommunication window at 1550 nm. The filter resolution is 10 nm FWHM [32-34]. Similar work has been presented in Ref. [35, 36] and with a resolution below 1 nm in Ref. [37, 38].

The main technological challenge is the flatness of the mirror suspended on a membrane. A wedge error between the top and the bottom mirror accumulates with multiple reflections in the cavity and deteriorates the wavelength resolution.

Blomberg *et al.* at VTT, Finland, presented a Fabry-Pérot filter tunable in the range from 4.0 to 4.5 μm with 0.1 μm resolution. It is integrated in a spectrometer system with black-body source and IR detector for infrared gas analysis [39].

Arrays of different fixed wavelength elements

Correia *et al.* reported on a single chip spectrometer in silicon consisting of an array of 16 photodiodes with different fixed wavelength filters on top. They

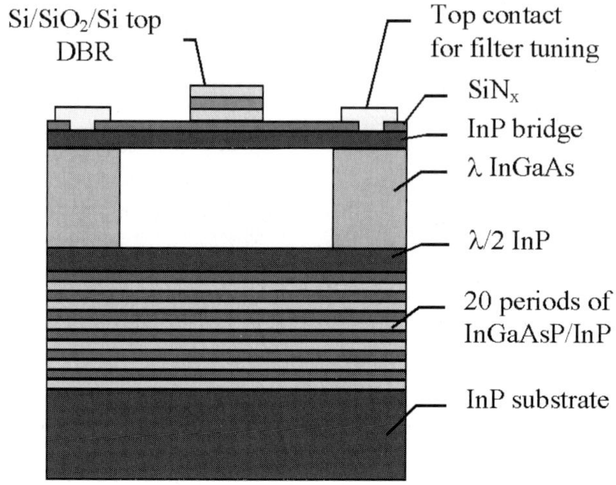

Fig. 1.12 Schematics of tunable Fabry-Pérot filter based on III-V semiconductor material. Courtesy of M. Strassner, KTH, Sweden.

cover the visible spectrum with 16 points that have a resolution of 16 nm FWHM [40].

Vertical cavity surface emitting lasers (VCSELs) can be tuned over some nanometers by the supply current. An array of 7 x 20 elements with slightly shifted central wavelengths can cover a range of 50 nm in the visible spectrum and has been proposed for sensing applications [41, 42].

Grating microspectrometers

Kiang *et al.* presented a grating on a polysilicon plate for spectroscopic applications [43]. The grating is mounted on torsion bars and turned by an electrostatic combdrive to change the optical incidence angle and tune the reflected wavelength picked up by a fixed point detector. The grating has a rectangular step modulation that causes also higher order diffraction modes. Phase gratings with a sawtooth profile could avoid this, but they are more difficult to produce.

Senturia, Hocker *et al.* presented a programmable reflective diffraction grating called polychromator [44-46]. It consists of 1024 grating-element actuators that can be controlled individually to adjust their optical phase. The system synthesizes parts of the infrared spectra of compounds by changing the configuration of the diffraction grating, which allows to determine rapidly unknown chemical samples by their infrared absorption bands.

Müller *et al.* developed a microspectrometer in LIGA technology (LIthographie, Galvanik, Abformung) [47-49]. An autofocussing phase grating projects a spectrum on a photodiode detector line (Fig. 1.13). The spectral resolution is 7 to 12 nm in the visible range. A near-infrared version (1.1 to 1.75 μm) exists of this device with an InGaAs detector line, which dominates the costs. A similar device called "Plane Concave Spectrometer" (PCS) for visible light has been developed at CSEM Neuchâtel, Switzerland, by Teichmann, Gale *et al.* [50]. Yee *et al.* report on a low cost spectrometer with 9 nm resolution in the visible range, consisting of a grating mounted on top of a CCD chip [51], similar to a two wafer device presented by Kwa *et al.* [52].

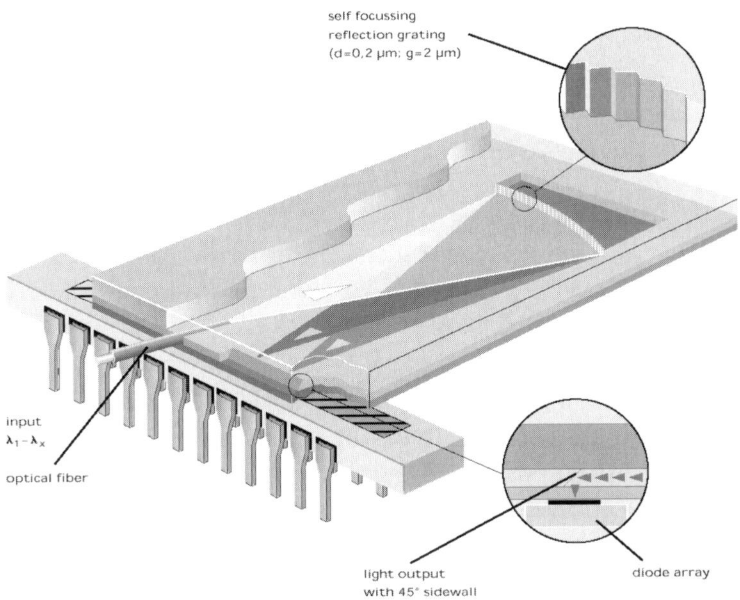

Fig. 1.13 Visible light LIGA spectrometer with autofocussing grating and photodiode detector array, developed by C. Müller. Courtesy of Steag Microparts GmbH, Dortmund, Germany.

1.3.7 Medical applications

Endoscopes allow surgeons to look inside the human body. The provided view is not good enough for tumor classification or similar complex diagnostics. Thus, the standard practice is to examine samples in a lab with confocal microscopes and fluorescent dyes, which can take from 30 minutes to several hours, while the patient and surgical team wait in the operating room.

Microdevices may soon change that. Confocal laser scanning microscopes have been miniaturized to the size of a fountain pen, able to fit into the tip of an endoscope. Several research groups [53, 54] are working on a miniaturized confocal microscope, making perhaps possible in the future to perform tissue diagnostics during the endoscopic examination.

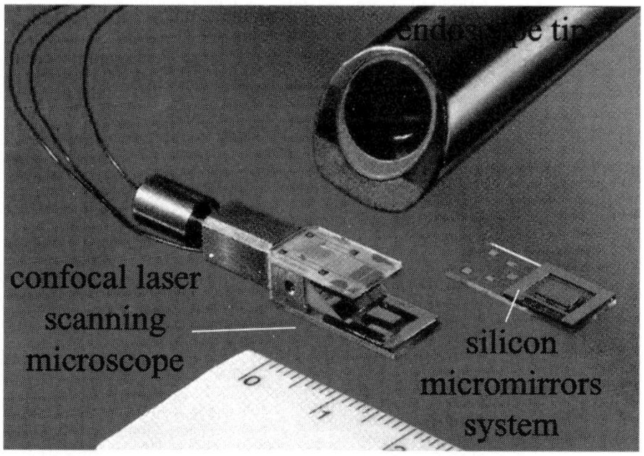

Fig. 1.14 The latest confocal laser scanning microscope of the Fraunhofer Institute of Itzehoe, Germany, fits into the tip of an endoscope. After Hofmann et al. [53].

Other applications may be relevant in a field of ophtalmology, called keratectomy, used to correct some types of ametropia and astigmatism. In this technique a laser must be positioned precisely at given points on the surface of the cornea and focused in depth despite the movement of the eye. A complete immobilization of the eye is needed, which can only be achieved by heavy narcotization. A micromachined scanning mirror as part of a keratotomy system should be small enough to be set directly on the eye ball and thus follow its movement [55].

1.4 MOEMS ACTUATION PRINCIPLES

Many optical beam manipulation devices based on MEMS technology were presented over the last twenty-five years. As the number of examples is impressively high, a non-exhaustive overview is given below. The following micro-scanners and light deflectors are classified by their actuation principles.

1.4.1 Electrostatic actuation

The most common actuation principle in the field of micro-deflectors is the electrostatic actuator. The simplest form of actuation for micromechanical light manipulators is electrostatic, and thus it represents one of the earliest approaches used despite its inherently nonlinear response to the control voltage. An other aspect is that the electrostatic driving forces are significant and efficient in the micro domain.

History

As for many micromachined devices, the origins of MOEMS beam manipulators date back decades before the terms "MEMS" or "micromachining" were known. Since the advent of the "Eidophor" projection light valve, which is an electrostatically deformed oil film on top of a reflector and is used for very large projection TV systems [56], various attempts have been made to replace this technology with a less complex one, using a number of other non-oil-film light valve systems for image projection [57].

Apparently, the earliest micromachined mechanical light modulator was the "mirror matrix tube" projection display device developed at Westinghouse Research Labs (Pittsburgh, PA) in 1975 [58, 59]. Their mirror array was deflected electrostatically relatively to the substrate when charged up by an electron beam. SiO_2 membranes grown on epitaxial silicon on sapphire, released by undercutting the silicon and coated with aluminum formed the deflectable mirror surfaces. Despite successful demonstration of the devices, they were never commercially viable, because of fabrication yields, reliability problems during the mirror free etching and the inability to achieve good enough electron-beam focusing on each individual pixel.

Two years later, Petersen [2] investigated silicon dioxide cantilevers for light modulators fabricated by an anisotropic silicon etch. The cantilevers could be individually deflected by applying a voltage to the capacitor-like structure. A 16-element light modulator linear array composed of the cantilevers was used in a small display system. A line of He-Ne laser light was reflected by the cantilever, apertured, focused and scanned by a galvanometer.

Torsional silicon light modulators

As an alternative optical deflector for display or galvanometer applications, silicon torsion mirrors were fabricated by K. Petersen at IBM based on a two-substrate construction [60, 61]. A glass substrate was etched to form a well into which the torsion plate could deflect. A support ridge was used to prevent it from being pulled down in parallel to the base of the well, and a pair of metal deflection electrodes were fabricated in the well (Fig. 1.15 a). For a mirror size of 2 x 2 mm, a resonance frequency of 15 kHz and a deflection angle of ±1° were achieved.

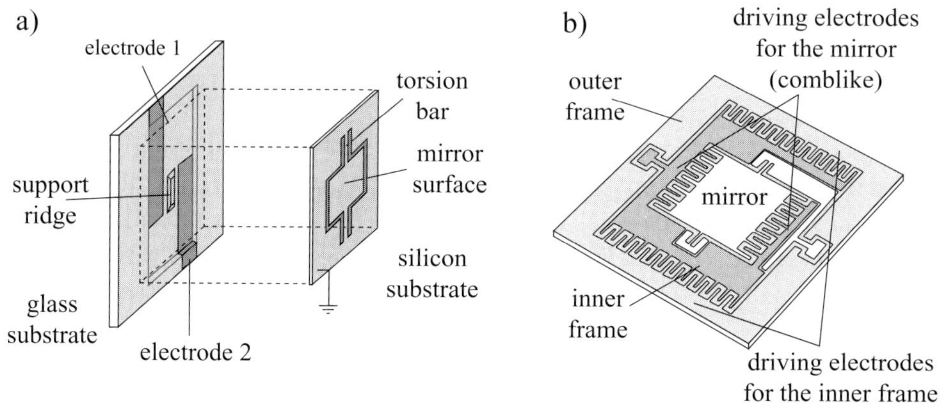

Fig. 1.15 a) Exploded view of Petersen's silicon torsion mirror structure showing the etched well, support ridge and the evaporated electrodes on the glass substrate [60].
b) The latest 2-dimensional silicon scanner from the Fraunhofer Institute in Dresden, Germany. After Schenk et al. (1999) [9].

Continuous operation of the torsional devices (near resonance) for months, followed by dislocation-sensitive silicon etching showed no damage. This highlights one of the important advantages of using single-crystal materials and of scaling down to these dimensions: fatigue mechanisms are not the same as those found in macroscopic materials due to the absence of grain boundaries.

More recently, the Technical University of Chemnitz, Germany, demonstrated the fabrication of single-crystal silicon torsional light deflectors as large as 3 x 3 mm using combinations of dry and wet etching [6, 62]. They also demonstrated the use of square silicon plates for 2D light beam scanning. These plates were designed with four underlying deflection electrodes, and were supported by compliant silicon members to allow semi-independent actuation in two orthogonal directions.

Similar work based on Petersen's torsional silicon mirror has been developed by various research groups [63-66]. However, for torsional mirror configurations where the driving electrodes are positioned behind the plate, the deflection angles are limited by the small gap between mirror and electrodes, whereas, for some applications, high deflection angles are required. The deflection angle can be increased by augmenting the gap, but in this case, large voltages become necessary.

The Fraunhofer Institute in Dresden, Germany, [67, 68] developed an innovative solution that avoids high voltages, which means that the control circuit can also be integrated on a silicon chip. A one-dimensional scanner version was built where two torsional springs suspend the silicon reflector plate in free space. The vertical sides of the lateral deposited driving electrodes and of the mirror plate form a variable capacitor, depending on the deflection angle of the suspended plate. The variation of the capacity during oscillation can be used to transfer energy into the oscillating system. To increase the possible feed-in of energy the capacity is increased by comblike electrode fingers. Scan angles of up to 40° can be achieved at less than 20 V, while the oscillation frequencies range from 140 Hz to 20 kHz and the mirror area from 0.5 x 0.5 mm up to 3 x 3 mm, respectively. A two dimensional scanner is in development (Fig. 1.15 b), where the mirror and the inner frame are excited independently [9]. Based on a SIMOX process, the mirror plate with the comb electrode are realized by deep dry silicon etching.

Torsional metal light modulators

A pioneering fabrication technique based on aluminum micromirrors released from an organic spacer layer was developed by Hornbeck *et al.* at Texas Instruments (TI). After the invention of the digitally operated micromirrors, this sophisticated approach was decisively followed to build up a whole system of mirror arrays and integrated driving circuits. The mirrors are fabricated from three metallizations, silicon dioxide and organic dielectrics which are processed on top of a 0.8 μm CMOS process. The 16 x 16 μm^2 aluminum mirror is suspended between posts on diagonally opposite corners of the square. The differential electrostatic attraction between the mirror and the electrodes causes the mirror to switch about the hinged diagonal so that the floating tip of the mirror touches down the landing electrode (Fig. 1.16 a) [69]. Since the rotation of such electrostatic devices is highly nonlinear with respect to the applied potential, it is advantageous to use them in a digital manner (full deflection one way or another, with little or no time spent in the undeflected state), hence the name "digital micromirror device" or DMD™. Current devices operate through ±10° angles with mechanical switching times on the order of 15 μs. Total deflection voltages are on the order of 20 V. Single-chip light valves of up to 2.3 million pixels were achieved and projection TV and HDTV displays are commercialized [3, 70] (section 1.3.1, Fig. 1.3).

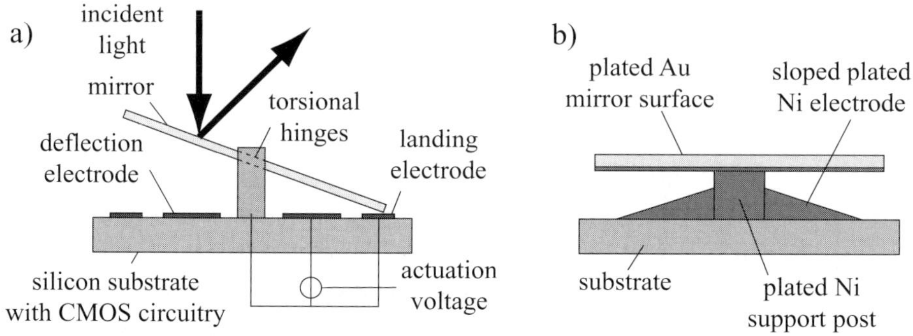

Fig. 1.16 a) Basic operation of a Texas Instruments torsional electrostatic micromirror. The tip of the plate is allowed to land on an equipotential "landing electrode" to prevent sticking. Current mirror designs have evolved considerably over time and are more complex [71].
b) Illustration of the electroplated reflective light modulator with sloped electrodes. Adapted from Wagner et al. [72].

Many other approaches for fabricating metal torsional micromirrors emerged from the work of TI. The Swiss Federal Institute of Technology in Zürich developed micromirror arrays consisting of the CMOS second metal layer deposited in two successive passes in order to establish a thick metal layer for the stiff mirror as well as a thin one for the flexible hinges [4, 73]. A metallic torsional mirror design was also presented by the Seoul National University in Korea. Their aluminum mirror (100 µm x 100 µm) is supported by electroplated nickel posts and is capable of large tilt angles of more than ±10° for an applied voltage of 40 V [74, 75]. The Fraunhofer Institute of Itzehoe in Germany (Wagner *et al.* [72]) worked on electroplated nickel mirrors having the same dimensions as the korean ones and supported by nickel posts as well (Fig. 1.16 b). A metal surface micromachining process has been developed using thick UV-resist, multiple electroplating, a copper sacrificial layer and a polishing process step. Nonplanar sloped electrodes reduced drastically the driving voltage of the electrostatic actuator. The mirrors operated over deflection angles of ±15° at 35 V threshold voltages.

Torsional polysilicon light modulators

During the last decade, low stress polycrystalline silicon (polysilicon) was developed for free standing MEMS structures, where silicon dioxide is used as a sacrificial layer. Polysilicon micromirrors of excellent mechanical stability were thus fabricated [76-79]. However, stress free polysilicon requires long annealing times at high temperatures, and metal interconnects and bond pads must be protected against the long HF sacrificial oxide etch.

Deformable grating light modulators

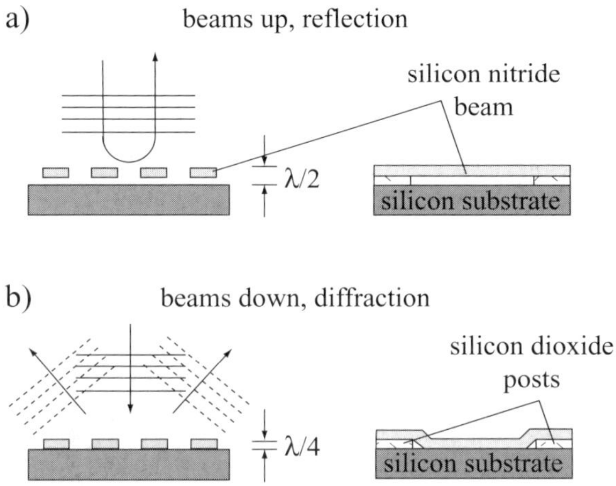

Fig. 1.17 Illustration of the operation of a deformable grating light modulator. After Solgaard et al. (1992) [79].

Besides the spatial light modulators, phase light modulators were fabricated using MEMS technologies [14, 80]. Deformable gratings were made of silicon nitride [79]. The grating depth was electrostatically controlled and the grating pitch designed appropriately to the wavelength of the light to be modulated (Fig. 1.17). A display with high color saturation can be built from a 2D array of deformable grating valve devices. By using three different grating periods, each of the three primary colors can be generated. Such displays are commercialized by Silicon Light Machines, Inc. (section 1.2.2, Fig. 1.7).

A different original work describes polysilicon micromotors with single and pyramidal diffraction grating elements fabricated on the polished surface of large-area rotors for optical scanning applications [81].

Membrane light modulator

Vdovin *et al.* at Delft University, The Netherlands, [15, 82] demonstrated a matrix display consisting of a large silicon nitride membrane glued on top of a driver chip. The membrane was covered by aluminum and electrostatically deflected at the location of individual pixels. Bifano *et al.* at Boston University [83] developed a single compliant optical polysilicon membrane supported by multiple attachments to an underlying array of surface normal electrostatic actuators. It is based on a three layer polysilicon surface micromachining technology.

The advantage of these types of structures is that they can be used to achieve a 100% pixel fill-factor, which is not possible for the other architectures. They have also the potential of low-cost manufacturing due to their simplicity. However, the membrane actuator shows large cross-talk between individual pixels and limited optical efficiency.

Integrated free-space optical systems

Surface micromachining, in which microoptical systems are batch-fabricated and placed on top of a silicon wafer, has become a promising approach to integrated free-space optical systems. With the demonstration that surface-micromachined elements can be "folded" out from the substrate, an important new degree of freedom has emerged. With this degree of freedom, original new optical structures based on electrostatic actuation could be fabricated, such as fresnel lenses, refractive lenses, micro beam splitters, Mach-Zender interferometers, micromirrors and micropositioners. The assembling of these components allowed the realization of complex systems such as microscanning systems, raster scanners and microoptical switches.

The research group of K.Y. Lau and R. S. Muller [84] at Berkeley Sensor & Actuator Center (BSAC), the group of M. C. Wu [85] of the University of California in Los Angeles (UCLA) in collaboration with L.Y. Lin from AT&T Research Lab, Sandia National Laboratory (NM) [86] as well as the Air Force Institute of Technology (OH, USA) with the group of V. M. Bright [87] developed free-space microphotonic systems using the well established polysilicon/sacrificial oxide technology. Mostly based on Pister's *et al.* [88] polysilicon hinge and sliding guides designs, the necessary out-of-plane orientation for the optical elements were achieved. Comb-drive actuators, linear microvibromotors [89] or microengines [85, 90] were used to move polysilicon torsional resonating micromirrors for sophisticated scanning applications (Fig. 1.5 and Fig. 1.8 a, section 1.3.1) [7, 8, 91].

1.4.2 Piezoelectric actuation

In the piezoelectric actuation, the electrically induced strain is approximately proportional to the applied electric field. Potential advantages include high stress, high bandwidth, high energy density, and potentially lower operating voltages than electrostatic actuation (depending on the geometry). Disadvantages include much smaller dimensional variation under drive than for electrostatics and sophisticated deposition processes. This explains why an external piezoelectric ceramic actuator is often assembled with a bulk silicon micromachined raster scanning system consisting of two mirrors [12, 18].

However, one-dimensional and two-dimensional micromirrors with integrated sol-gel deposited thin film PZT bimorph actuators have been reported by Schroth *et al.* [92]. Depending on the width of the beam, maximal scanning angles of 35° could be measured at an applied AC voltage of only 5 V_{pp} at a resonance frequency of 24 kHz. Ohtuka *et al.* [93] developed a similarly actuated two-dimensional scanner integrated in a planar structure and based on a torsional mirror. ZnO thin film has also been used successfully in mirror array application [94].

1.4.3 Magnetic actuation

Magnetic actuation has many advantages over both electrostatic and piezoelectric actuation. For example, magnetic forces in a microactuator can be both attractive and repulsive. They can also be large in magnitude (on the order of milliNewtons) and effective over a long range (on the order of millimeters). The major disadvantage of magnetic actuation is the lack of a mature technology. This is mainly because magnetic actuation requires more complicated components, such as coils and magnetic materials.

Miller *et al.* from Caltech, Pasadena, CA, demonstrated an electromagnetally actuated scanning mirror developed for holographic data storage. An electromagnetic coil as well as a Permalloy magnetic layer (FeNi) are integrated on a thin cantilever beam (Fig. 1.18 a). When an external magnetic field is applied and there is no current in the coil, the mirror deflects to a "bias" angle, due to the Permalloy only. Current supplied in the coil allows for movement about this bias position. Using a high external field to saturate the Permalloy (994 Oe), the bias angle was 58°, with ±10° deflection with moderate linearity. The first resonant mode of the structure was at 65 Hz [13, 95]. Miller also demonstrated an electromagnetic fiber optic 2 x 2 light microswitch [96].

A different design consists of an integrated planar coil on a thin silicon mirror (Fig. 1.18 b) [97]. Immersed in a external magnetic field, the mirror is able to rotate around suspended torsion bars, deflected by the Lorentz force. This design has no transverse force acting on the rotation axis during scanning, which provides good mechanical characteristics as low distortion and damping, high optical angular deflection (≥22°) at a resonance frequency of 1.5 kHz. Based on the same torsional silicon mirror design, Olympus Optical Co., Ltd., Japan [98]

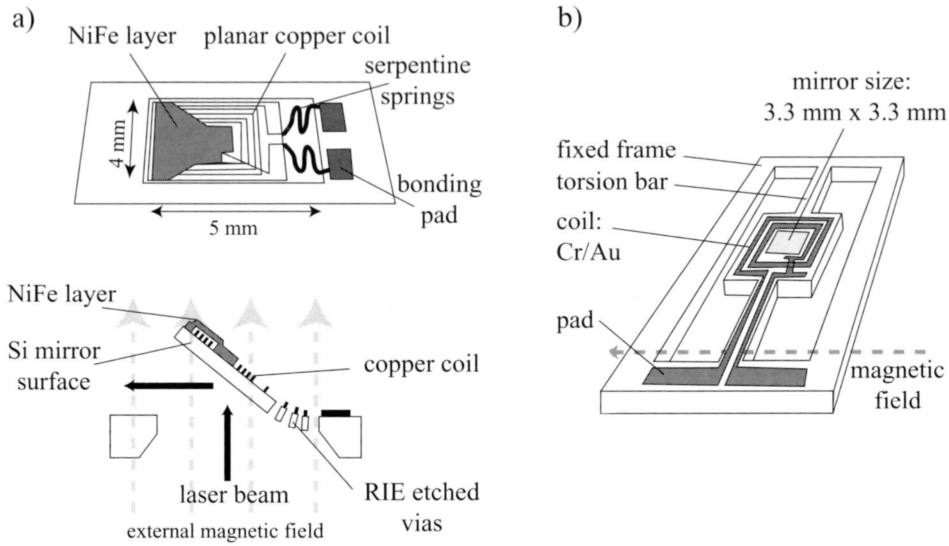

Fig. 1.18 a) Schematic top view and cross-sectional view (mirror deflected) of Miller's microscanner. Adapted from Miller et al. [13].
b) Schematic drawing of Ferreira's silicon scanner structure. After Ferreira et al. [97].

worked on optical microscanners capable of scan angle control with the integration of a sensing coil and newly developed polyimide hinge structures, allowing good durability and shock resistance.

In addition to that, many developments in the field of magnetostrictive materials have shown that magnetic materials can still produce high enough force in the micro / millimeter range to build a large variety of micro-actuators with innovative applications [99]. In this context, a micro-resonator for two-dimensional light deflection applications was presented [100].

1.4.4 Thermal actuation

Thermal and thermal bimorph actuators provide greater force and larger deflections than similarly sized electrostatic or piezoelectric actuators. They have the advantage of nearly linear deflection-versus-power relationships, they are often based on simple fabrication steps and their structure allows for easy integration. Disadvantages include high power and low bandwidth (determined by thermal time constants).

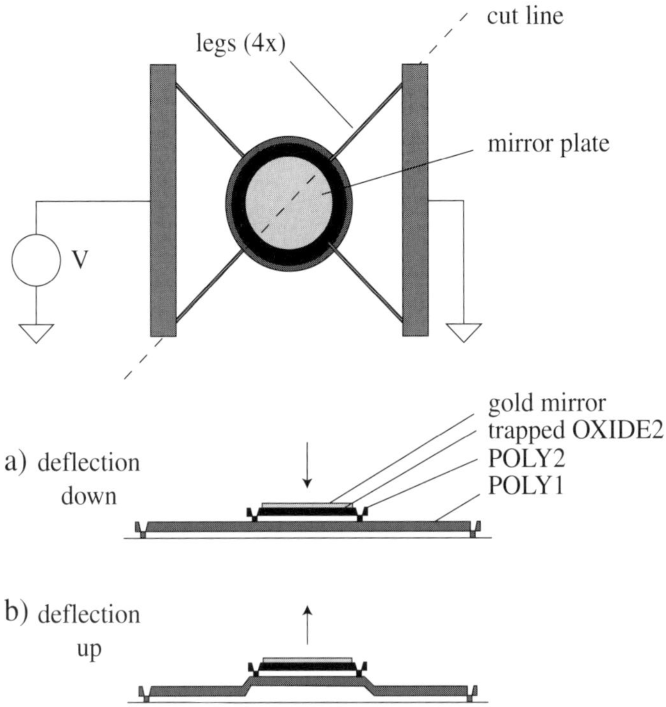

Fig. 1.19 Schematic view of basic thermal piston micromirror fabricated in a polysilicon surface micromachining process. After Cowan et al. [101].

A biaxial scanning mirror activated by bimorph (silicon / aluminum) structures for medical applications has been previously presented [55] (section 1.3.7). However, this device suffered from large power dissipation and slow response behaviour. Bühler *et al.* from the Swiss Federal Institute of Technology in Zürich showed that thermally actuated micromirrors with integrated on-chip circuitry can be easily fabricated using a standard CMOS process. Their bimorph cantilevers were composed of aluminum and silicon dioxide [102].

The Air Force Institute of Technology (OH, USA) with the group of V. M. Bright developed thermally actuated piston micromirror arrays [101]. Their micromirrors offered large deflections at drive voltages compatible with CMOS circuitry. Fig. 1.19 shows schematically a top view and diagonal cross section of the piston micromirror. When current passes through the polysilicon legs, resistive heating induces thermal expansion. As the legs increase in length, the mirror is forced to deflect. Maximal deflection of more than 2 μm at 120 mW was measured. Operation at 1 kHz is feasible. Initial tests indicate that thermal piston micromirrors may be an attractive alternative to electrostatic designs for optical phase modulation applications.

The Rockwell Science Center (CA, USA) developed a one-dimensional scanner consisting of a thermal bimorph actuator and a mirror on top of a silicon cantilever beam (Fig. 1.20) [16, 103]. The actuator consists of ZnO between two metal layers, which function as top and bottom electrodes. The mirror has an area of 12 mm^2. A maximal scanning angle of 20° for a driving voltage of 1 V and at a resonance frequency of 300 Hz was achieved.

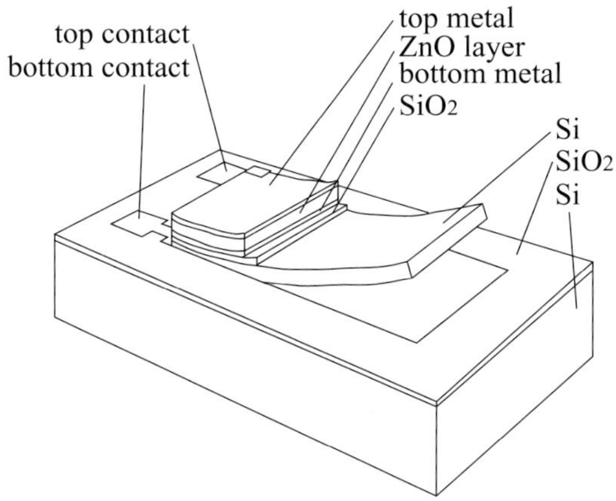

Fig. 1.20 Schematic of a typical Rockwell scanner profile. Adapted from Motamedi et al. [16].

2 BASICS FOR A THERMALLY ACTUATED MICROMIRROR

There is nothing more practical than a good theory.

2.1 MICROACTUATOR SPECIFICATIONS

The following microactuator principle will be used for various applications: a miniature scanning device for barcode reading, a microprojector and a tunable filter in a microspectrometer. The specifications are similar.

The scanner should show high scanning deflections, in the order of 50° optical scan amplitude. Therefore, a highly efficient actuator is required. Its high amplitude deformation must be obtained in a frequency driving range from 50 to 500 Hz. The actuator has to be capable of a reasonably fast response time. Low power consumption (in comparison to macro-devices) is desired for eventual hand-held applications. Simple technology and standard materials are necessary to guarantee a low cost batch fabrication process and a final reliable inexpensive device.

A thermal actuator was chosen due to the large forces and the large deflections it can provide (see chapter 1, section 1.4). In addition, the simplicity of such structures allows easy integration and a simple fabrication process. A disadvantage of thermal actuation includes high power consumption compared to other microactuators. Nevertheless, the consumption is sufficiently low compared to current macro-devices. Low bandwidth (determined by thermal time constants) is an other disadvantage. However, as shown subsequently, the device can be excited at its mechanical resonance frequency which is reasonably high for such microstructures and acceptable for the specified application.

Chapter 2 and 3 are strongly interrelated. Some understanding of the fabrication technology is needed to fully understand this chapter. Details about the fabrication process are given in Chapter 3. A special fabrication process using different materials for a thermal actuator based on porous silicon will be presented in chapter 7 in the context of a microspectrometer. The basic considerations and calculations apply to all of these devices.

2.2 PRINCIPLE OF THE PRESENTED MICROSCANNER

For reasons of simplicity and as the two dimensional microscanner does not differ much from the one dimensional micro-mirror, Chapter 2 is focused on the latter only.

The layout for the simple design consists of a mirror attached to the end of one or two *bimorph cantilever beams*, which are thermally actuated (Fig. 2.1). The single beam layout incorporates the electrical path on one beam; whereas for the double beam configuration, the electrical path goes forth on one beam and back on the other. The two components of the bimorph layer are a metallic film (chromium) and SiO_2. The cantilever part of the device is naturally bent upwards, due to residual compressive stress in the oxide and residual tensile stress in the evaporated metal layer. This property is exploited in order to achieve an *out of plane rest position* θ_{in} ('*in*' is for initial), which permits high scanning amplitudes and allows for hybrid optical components (laser diode, lenses) to be placed directly on the substrate.

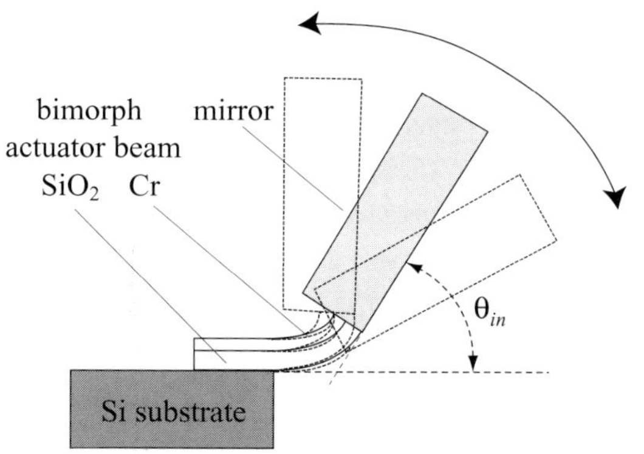

Fig. 2.1 Principle of the scanning unit.

2.3 THE THERMAL BIMORPH ACTUATOR

A sandwich of two layers of materials is called a bimorph structure. These actuators convert small strain differences between layers to relatively large displacements perpendicular to the strain plane. The bimorph, free standing cantilever can be heated by an electric current through the conductive layer. The different coefficients of thermal expansion of the composing layers lead to bending of the beam.

2.3.1 Thermal expansion

Thermal expansion of materials results generally from the asymmetry of the potential the atoms in a solid are subjected to. This potential is called the Lennard-Jones-Potential and is composed of a repulsive and an attractive contribution. By increasing the temperature, the distance between the equilibrium positions of the neighboured atoms becomes usually higher. Therefore, the thermal expansion coefficient is in most cases positive.

For a free beam or a free material layer, the strain created by thermal expansion in an uniaxial consideration, thus a scalar notation, can be expressed as follows:

$$\varepsilon = \alpha \cdot \Delta T \tag{2.1}$$

where ε is the strain, α (K^{-1}) the thermal expansion coefficient and ΔT (K) the temperature variation.

For a double clamped beam or an unreleased material layer, inserting the scalar notation of Hook's Law ($\sigma = E \cdot \varepsilon$), the thermal stress becomes:

$$\sigma = E \cdot \alpha \cdot \Delta T \tag{2.2}$$

with σ (Pa), the stress and E (Pa) the Young's modulus. Eqn. 2.2 shows that the stress generated by thermal expansion is independent of the dimensions of the device. Therefore, the corresponding force scales with square of the size as far as the material withstands the thermal stress. The thermal force gains over inertial forces as the size of the system is decreased.

The stresses and with them the forces that can be generated by thermal expansion are large whereas the displacements are small. Therefore, a mechanism amplifying the displacement is necessary, for example, by a lever mechanism, by buckling or by using a bimorph structure.

2.3.2 Stress in bimorph cantilevers and initial rest position

The cantilever of the device is naturally curled up, due to residual stresses in the different components of the bimorph cantilever. Thus, the device has a given rest position. For our applications, it is important to determine the angle θ_{in} (Fig. 2.1).

Stress in thin films

Nearly all films are found to be in a state of residual internal stress, regardless of the means by which they have been produced. The stress may be compressive or tensile. Compressively stressed films tend to expand parallel to the substrate surface. Films in tensile stress, on the other hand, contract parallel to the substrate.

The total stress σ in a thin film is the sum of any extrinsic stress σ_{ext} on the film and the residual stress σ^*:

$$\sigma = \underbrace{\sigma_{th1} + \sigma_e}_{\sigma_{ext}} + \underbrace{\sigma_{th2} + \sigma_i}_{\sigma^*} \tag{2.3}$$

Extrinsic stresses are composed of external stresses σ_e which are imposed by external factors such as packaging induced stresses or stress from another film and thermal stresses σ_{th1} resulting from the difference in the coefficients of thermal expansion between the film and the substrate. Residual stresses are composed of thermal stresses σ_{th2} as well, resulting from deposition temperature conditions and intrinsic stresses σ_i developed during the film nucleation. Intrinsic stresses are not yet completely understood. Several models relating to the origins of the intrinsic stress have been proposed. These invoke the following stress causing mechanisms: lattice mismatch between the substrate and the film, rapid film growth (which locks in defects), recrystallizing process, phase transformations and incorporation of impurities into the film [107, 121, 122].

Measuring stress in thin films

The disk method, most commonly used for stress determination in thin films, is based on the measurement of the change of curvature on a wafer substrate caused by a stressed film. A laser scan over the surface and determination of the reflection angle allows to calculate the stress from the measured radius r. This is done before and after film deposition.

The Stoney equation (Eqn. 2.4) [121] allows the determination of the deduced film stress σ_f from the measured radius, asserting the following assumptions:

- the disk substrate is thin and has transversely isotropic elastic properties with respect to the film normal,
- the applied film thickness is much less than the substrate thickness ($t_f << t_s$),
- the film thickness t_f is uniform,
- the temperature of the disk substrate / film is uniform,
- the disk substrate / film is mechanically free,
- the stress is equi-biaxial and homogeneous over the entire substrate,
- the film stress is constant through the film thickness.

The stress in the film can be written as:

$$\sigma_f = \frac{E_s}{6 \cdot (1 - \nu_s)} \frac{t_s^2}{t_f} \left(\frac{1}{r_{sf}} - \frac{1}{r_s} \right) \tag{2.4}$$

where E_s is the elastic modulus of the substrate, ν_s the Poisson's ratio of the substrate, t_s and t_f respectively the substrate and the film thickness, r_s and r_{sf} respectively the curvature radius of the substrate without film (*s*) and with the film to be measured (*sf*).

For the determination of the initial curvature of the device, the evaluation of the residual stress in the two deposited layers is necessary. Bimorph cantilever components internal stresses (thermal wet grown silicon dioxide layer and evaporated gold/chromium layer) have been measured with the disk method.

The wet thermal grown silicon dioxide is a compressive layer, consequently, the stress is negative. Metal thin films are usually tensile and the evaporated Cr layer is a tensile layer. The stress values found in the literature for evaporated films should be considered representative rather than precise. Published data is often inconsistent, even if the investigators employed similar measurement techniques. Indeed, it has been observed that in general the intrinsic stress in a film depends, among other parameters, on thickness, deposition rate, deposition temperature, ambient pressure, method of film preparation, type of substrates used and type of machine [122]. These aspects justify our measurements.

The main source of error for this type of measurements is the positioning change of the sample before and after the film deposition. The induced error in the radius measurement will introduce a standard deviation of about 40 MPa in the calculated stress.

Table 2.1 Average stress of the bimorph layers for a wafer batch, processed at the same time.

	Thickness of the layer t [μm]	Residual stress σ^* [MPa]
SiO_2	1	-332 ± 8
Au/Cr	0.4 (0.03 Au, 0.37 Cr)	325 ± 14

Note that the gold layer is just a thin 30 nm film, used as seed layer for the mirror electrodeposition (Chapter 3). The bilayer Au/Cr is considered as a unilayer. The measurements have been done on a wafer batch, processed all together. The stress variation between two samples is smaller than the error measurement due to the positioning (Table 2.1). The stress value of SiO_2 is well known (300…400 MPa) and our measurements are close to the data found in the literature [122].

Curvature due to residual stress in thin films of a bimorph beam

A simple analytical model is presented, allowing the determination of the rest position curvature of the curled-up bimorph beam, assuming linear elastic behavior of the materials (Hook's Law). Besides that, our model is based on the following assumptions:

- the beam considered is assumed to be symmetric about the *xy* plane (Fig. 2.2 a),
- the width of the beam layers is considered to be equal for both layers,
- the layer thickness is assumed to be homogeneous (Fig. 2.2 a),
- materials are considered to be isotropic and continuously distributed,
- $\sigma_z = 0$,
- the radius of curvature *r* is considered constant along the beam.

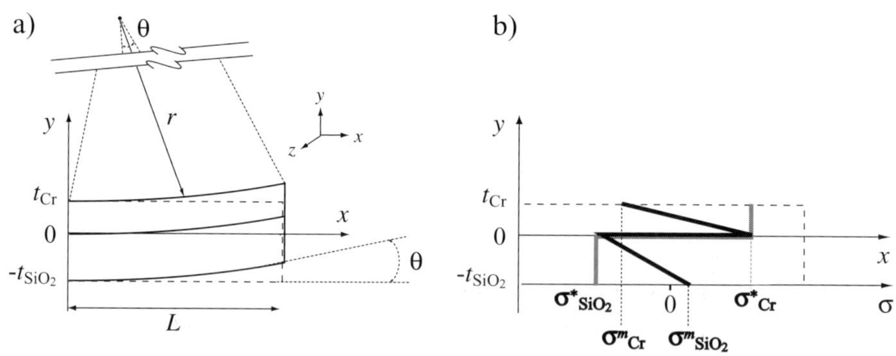

Fig. 2.2 a) Schematic drawing of the bimorph beam when released from the substrate. The interface of the two layer is denoted by y = 0. Note that, by convention, the radius of curvature r > 0 if the beam bends as depicted. t_{Cr} and t_{SiO2} are chromium thickness and silicon dioxide thickness, respectively. b) Stress representation in the bimorph beam before (grey) and after (black) the release of the beam.

Fig. 2.2 b) represents the stress distribution in the composed beam. σ^*_{Cr} and σ^*_{SiO2} stand for the residual stress in the Cr layer and in the SiO_2 layer. They are considered to be homogeneous over the thickness of the layer. σ^m_{Cr} and σ^m_{SiO2} are the stress at the end of the Cr layer ($y = t_{Cr}$) and in the SiO_2 layer ($y = -t_{SiO2}$), respectively, once the beam is released from the substrate. Due to the fact that the Cr layer, as the upper layer, is in tensile stress and tends to shorten after release while the compressed SiO_2 layer, as the lower layer of the composite cantilever, tends to elongate after release, the cantilever curls up (y positive) after removal of the substrate (Fig. 2.2 a). The Cr layer will be partially compressed by the bending: the residual tensile stress σ^*_{Cr} will progressively, over the thickness, become compressed. A compressive maximal stress σ^m_{Cr} is expected at the surface. In the same way, a tensile extremity stress σ^m_{SiO2} is expected.

The strain (in x-direction) of the layer of a bent cantilever can be expressed by the sum of (Eqn. 2.5):

- the mean strain $\varepsilon^*_{\mathrm{layer}}$ due to the process conditions,
- the $\Delta\varepsilon$ strain variation when the beam is released (which is the same in the two layers),
- the strain due to the bending moment and the flexural rigidity: y / r.

$$\varepsilon_{layer} = \varepsilon^*_{layer} + \Delta\varepsilon - \frac{y}{r} \tag{2.5}$$

Note that the bending strain is negative because for $y > 0$ (Cr), the moment tends to decrease the strain in Cr layer. The linear elastic condition leads to the expression of the normal stress (in x-direction) in the layer by:

$$\sigma_{layer} = \varepsilon_{layer} \cdot \frac{E_{layer}}{1 - \nu_{layer}} = \varepsilon_{layer} \cdot E'_{layer} \tag{2.6}$$

where E_{layer} and ν_{layer} are the elastic modulus and the Poissons's ratio, respectively. E'_{layer} is the biaxial elastic modulus.

The equilibrium in the released cantilever is reached when the total forces ΣF and the total bending moments ΣM are equal to zero (index Cr for chromium and index SiO_2 for silicon dioxide):

$$\sum F = \int_{-t_{SiO_2}}^{0} (\sigma_{SiO_2} \cdot dy) + \int_{0}^{t_{Cr}} (\sigma_{Cr} \cdot dy) = 0 \tag{2.7}$$

$$\sum M = \int_{-t_{SiO_2}}^{0} (\sigma_{SiO_2} \cdot y \cdot dy) + \int_{0}^{t_{Cr}} (\sigma_{Cr} \cdot y \cdot dy) = 0 \tag{2.8}$$

From this system of two linear equation, r can be extracted. After simplification (Appendix A.1), $1 / r_{in}$, the initial curvature can be expressed as follows:

$$\frac{1}{r_{in}} = \frac{6 \cdot (\varepsilon^*_{Cr} - \varepsilon^*_{SiO_2}) \cdot t}{\frac{E'_{SiO_2} \cdot t_{SiO_2}^3}{E'_{Cr} \cdot t_{Cr}} + \frac{E'_{Cr} \cdot t_{Cr}^3}{E'_{SiO_2} \cdot t_{SiO_2}} + 6 \cdot t_{SiO_2} \cdot t_{Cr} + 4 \cdot t_{SiO_2}^2 + 4 \cdot t_{Cr}^2} \tag{2.9}$$

let K denote:

$$K = \frac{6 \cdot t}{\frac{E'_{SiO_2} \cdot t_{SiO_2}^3}{E'_{Cr} \cdot t_{Cr}} + \frac{E'_{Cr} \cdot t_{Cr}^3}{E'_{SiO_2} \cdot t_{SiO_2}} + 6 \cdot t_{SiO_2} \cdot t_{Cr} + 4 \cdot t_{SiO_2}^2 + 4 \cdot t_{Cr}^2} \tag{2.10}$$

then

$$\frac{1}{r_{in}} = K \cdot \Delta\varepsilon^* \tag{2.11}$$

where

- $\Delta\varepsilon^*$ is the linear strain difference caused by residual stress, and
- $t = t_{Cr} + t_{SiO2}$.

Eqn. 2.9 confirms the positive contribution of the initial strain in the silicon dioxide layer which is added to the strain in the chromium layer.

As seen later in section 2.3.3, Eqn. 2.9 is similar to the curvature given for the thermal bimorph actuation and it can be shown that the initial curvature can be added to the curvature due to the thermal contribution.

Influence of the beam thickness

Eqn. 2.9 shows that the curvature of the beam is strongly dependent of the ratio of layer thickness of the cantilever. For given constant material parameters and for a given fixed SiO_2 thickness (only t_{Cr} varies), the highest curvature is obtained if the ratio between the thickness of the two layers is inversely proportional to the ratio of the biaxial elastic moduli (Eqn. 2.12) [124].

$$\frac{t_{SiO_2}}{t_{Cr}} = \frac{E'_{Cr}}{E'_{SiO_2}} \tag{2.12}$$

This characteristic is also valid and important for the design of the bimorph sensitivity. The sensitivity for a bimorph actuator is defined as being the deflection angle the cantilever is capable to bend per degree Kelvin.

If the thickness of the layers composing the bimorph actuator are adequately chosen, the angular deflection during actuation can be maximized and the rest position curvature will be maximized as well. This can be a constraint in the design for the initial position because a maximum initial curvature is not always wanted. Fig. 2.3 shows the curvature variation versus ratio between the thickness of the two layers for Cr initial residual stresses (σ^*_{Cr}). All the curves have a maximum for a given t_{Cr} / t_{SiO2}.

An optimum must be found between high bimorph sensitivity and a reasonable initial curvature position corresponding to the application specifications. To guarantee a reproducible initial curvature, it is important to have a well controlled deposition process to guarantee thickness homogeneity. This is quite well controlled in micromachining deposition techniques. For evaporation, a thickness homogeneity error of less than 5 % is certified.

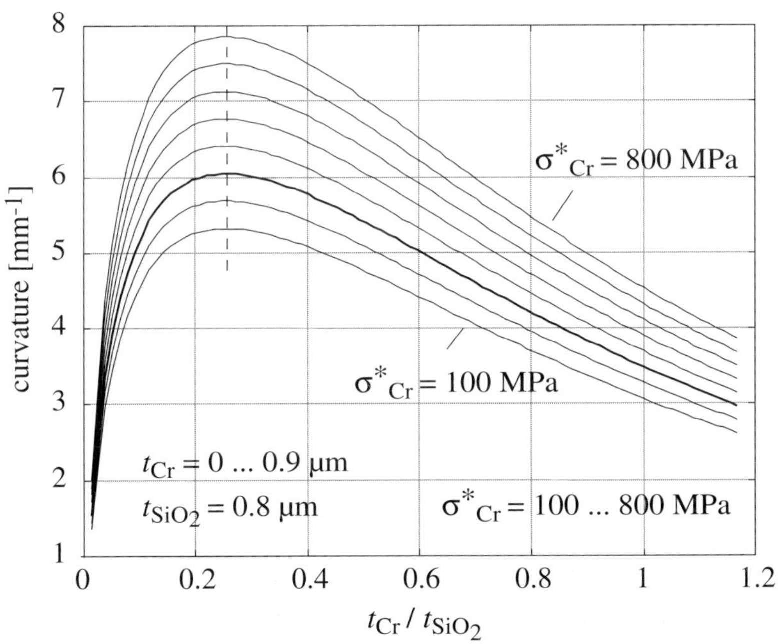

Fig. 2.3 Initial curvature vs. thickness ratio of the two layers for various Cr initial residual stresses (PVD Cr can easily have an initial stress of about 800 MPa, depending on parameter process [121, 123]). The bold line represents our initial stress conditions (σ^*_{Cr} = 325 MPa). The curves were plotted with the parameter in Table 2.2. A maximum is found for t_{Cr} / t_{SiO2} = 0.27.

Influence of material parameters

- *Influence of the residual stress in the thin films*

The initial residual strain deduced from stress measurements influences linearly the curvature (Eqn. 2.9) ($\sigma = E \cdot \varepsilon$). As the standard deviation of the measurement was about 40 MPa (15%), the curvature can be predicted with a 10% error, if the stress was measured for this specific wafer batch. In addition the uncertainty on the biaxial modulus has to be taken in account, as developed below. This last deviation is difficult to predict. In our case, the greater the absolute initial stress in the two chosen layers, the higher the curvature (Fig. 2.3). As reported in the paragraph on stress measurement, the residual stress in silicon dioxide, σ^*_{SiO2}, is a reproducible stable feature due to the well controlled process conditions. Therefore, the stress in the Cr layer is more susceptible to vary and may be an important issue in the curvature control.

- *Influence of the biaxial moduli*

The Young's modulus and the Poisson's ratio are parameters that are not always well known for thin films. They are often different from the bulk material, depending strongly on film quality and microstructure, thus on the deposition conditions. As for thermal wet grown SiO_2 the values are well known [125, 126], evaluating the mechanical parameters for evaporated chromium is already more difficult. The evaporation can be done at different substrate temperatures, at various deposition rates and at varying evaporation power. In the literature, we found Young's modulus values for evaporated PVD chromium varying between 100 and 280 GPa [127, 128].

For our devices, the Young's modulus of the Cr was measured by nano-indentation at the Université de Franche-Comte in Besançon, France [126]. The value is about 280 GPa (the same value as for the bulk, Table 2.2) and was measured with a precision of about ±10 GPa.

Fig. 2.4 shows the dependence of the curvature vs. the E_{Cr} moduli and the Poisson's ratio ν_{Cr} for the conditions reported in Table 2.2. One can see that the curvature decreases with increasing stiffness and lateral expansion. Actually, if the value $E_{Cr} = 280 \pm 10$ GPa is considered, a deviation in the curvature of ±0.7% is calculated. Despite the imprecisions on the determination of the mechanical characteristics, especially for the chromium layer, the error on the curvature determination remains small. For the remaining calculation, $E_{Cr} = 280 \pm 10$ GPa is used.

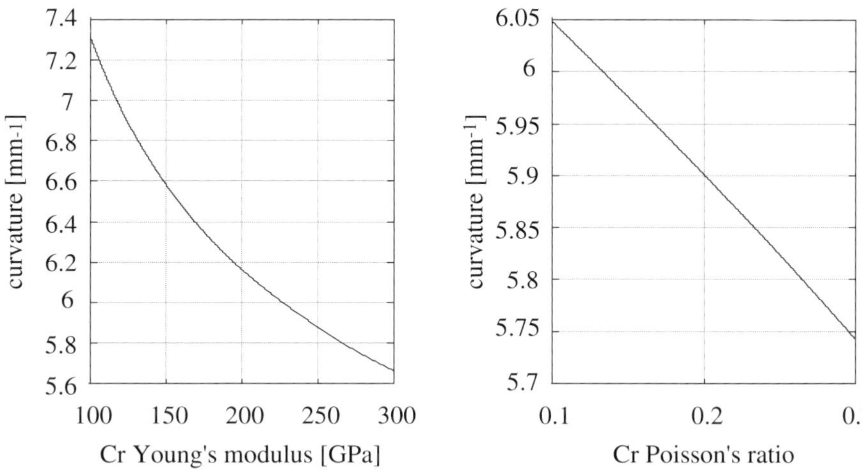

Fig. 2.4 Initial curvature vs. E_{Cr} and ν_{Cr}.

Comparison between measured and calculated rest position angles

Fig. 2.5 a) represents the stress distribution in the two layers at the rest position. The material parameters are reported in Table 2.2. As expected, the tensile Cr stress decreases to a compressive maximal stress along the thickness. A tensile extremity stress for the silicon dioxide layer is calculated (Table 2.2). A radius of curvature of 0.178 mm is deduced, thus a curvature of 5.6 mm^{-1}.

*Table 2.2 Mechanical properties and parameters used for the numerical calculation [126, 129, 4]. Bulk material values indicated by * are given where thin film data are not available [128]. The calculated stress at the end of the layers is given, too.*

	Young's modulus E [GPa]	Poisson's ratio ν	Residual stress σ^* [MPa]	Layer thickness t [µm]	Calculated stress at the end of the layer σ^m[MPa]
SiO_2	75 ± 5	0.17	-332 ± 8	0.8 ± 0.1	132.7
Cr	280 ± 10	0.3*	325 ± 14	0.4 ± 0.02	-308.5

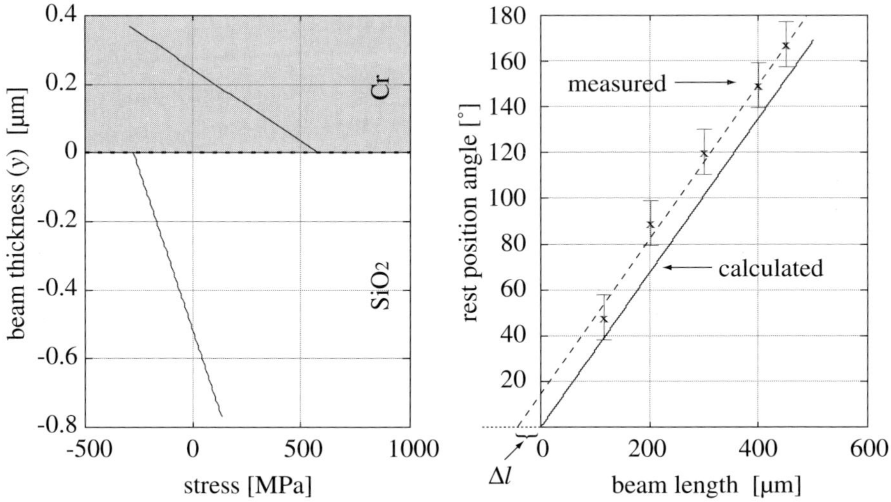

Fig. 2.5 a) Calculated stress distribution at the rest position in the microbeam. b) Rest position angle versus microbeam length, calculated for a curvature of 5.6 mm^{-1} (Eqn. 2.13, line) and measured (crosses).

From the initial value of the radius of the released beam r_{in}, the rest position angle θ_{in} can be calculated. θ_{in} is the angle between the substrate and the mirror (Fig. 2.1). L is the length of the actuator beam. The length of the deflection curve is equal to the initial length L because axial changes in length due to direct tension is disregarded. As the cantilever is unloaded, Eqn. 2.13 is valid for the assumptions made for the curvature calculation.

$$\theta_{in} = \frac{L}{r_{in}} \tag{2.13}$$

Eqn. 2.13 is plotted in Fig. 2.5 b for the given parameters (Table 2.2) and represented as a function of the beam length L. The observed initial angle for structures of different beam length is also depicted. The measured values were obtained by visual observation. They are the average result made on several wafers processed at different moments and with changing processing conditions, hence the deviation of ±10°.

The calculated angle is about 39.6° for a 120 µm long beam. For these dimensions, we observe an angle of about 45° ± 10° (Fig. 2.5 b) on processed structures. For a beam length above 300 µm, the structure bends more than 90°.

The measured values fit a linear dependence to the beam length (Fig. 2.5 b) with a similar slope as the calculated line. Indeed, the linear approximation of the measured values results in a slope of about 6.1 mm^{-1} (instead of the calculated 5.6 mm^{-1}). This is a good correlation and the small deviation can be explained by the imprecise material parameters that were used (E modulus, Poisson's ratio, initial residual stress), in particular regarding the evaporated chromium. The most probable cause of imprecision comes from the thickness of the silicon dioxide and the assumed constant thickness condition. As a matter of fact, t_{SiO2} will depend on the very last step of the process, where the scanners are released from the substrate in a KOH solution (Chapter 3). As this solution is not absolutely selective to the SiO_2 layer, the oxide is partially etched (Si / SiO_2 etch ratio of about 250) and its thickness is inhomogeneous (Fig. 2.6). We estimate an effective oxide thickness of 0.8 ± 0.1 µm (Table 2.2).

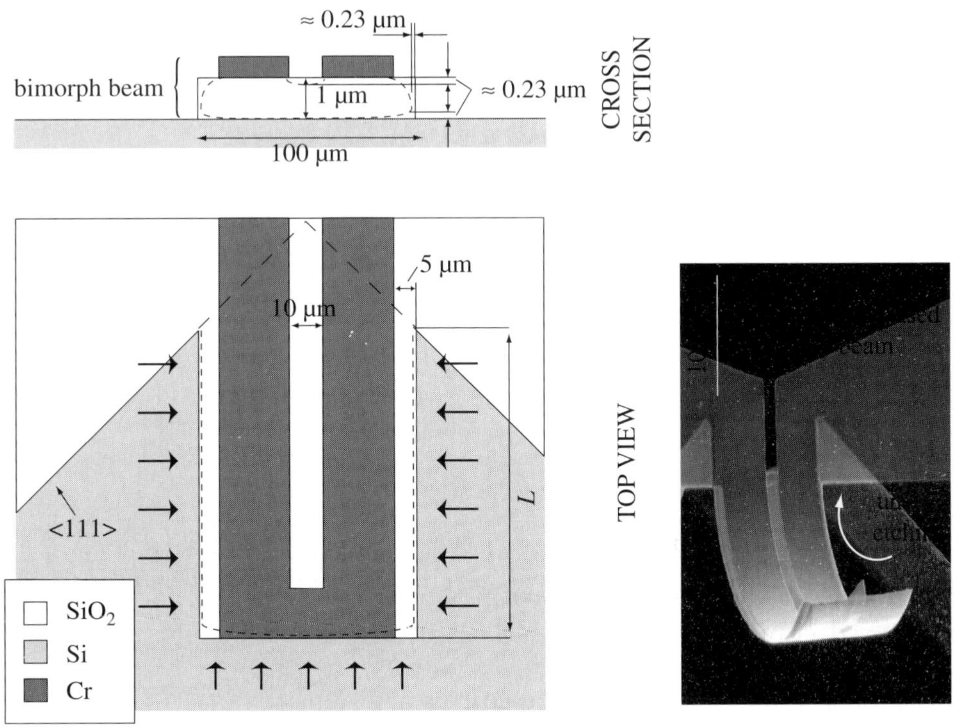

Fig. 2.6 Schematic and SEM picture of bimorph beam before and after the release etch. The dashed line represents the estimated shape of the SiO_2 layer after the KOH etch. Note that the mirror is not represented here and that the drawing is at scale.

A beam length offset Δl is observed in Fig. 2.5 b which could correspond to the neglected clamping conditions. Indeed, the length L is defined as represented in Fig. 2.6 and the underetched part is not considered in the calculations.

Rest angle position reproducibility

A collimated HeNe laser beam with a diameter of 1.2 mm is reflected by the surface of the mirror on a screen and by the substrate which provides a reference. The position of these two references provides, besides other informations, the rest position angle or the idle angle of the mirror. The idle angle is calculated from the average of the vertical spot positions. The measurements were performed by the IRO, EIV (Ecole d'Ingénieurs du Valais), Switzerland [130, 131].

The measurements have been performed for two different wafers. The mirrors have all the same geometry, but different bimorph thicknesses from wafer to wafer and the wafers were processed in two different cleanrooms. The repeatability on one wafer is good as all angles fell in a range of ±3.2 degrees (Table 2.3).

Table 2.3 Measured and calculated idle angles for 300 μm long beam bimorph beam.

	Cr layer thickness [nm]	θ_{in} measured mean value [°]	θ_{in} calculated [°]
wafer 1	450	129 ± 1.8	91.6
wafer 2	350	148 ± 3.2	100.2

Conclusions

This section puts attention to the important parameters acting on the initial curvature of a bimorph beam. The issues are residual stress in the layers due to deposition conditions, mechanical material constants which are not always well determined in thin films and the thicknesses of the two bimorph components. Measured values are close to the calculations. The repeatability of the measured initial angle is very good. In the section 2.5, it will be shown that when the bimorph is actuated, the initial angle will change due to a preliminary thermomechanical "burn-in" phase.

2.3.3 Thermal bimorph actuator design

Curvature due to thermal stress

Analytical solutions for tip deflection and curvature of cantilever bimorph beam actuators, consisting of homogeneous layers, have been presented in many papers [124, 129, 4, 132, 133, 134].

It can be easily shown that the two curvatures (the initial and the thermal one) are additive [4] (Eqn. 2.9). The final bending radius r_{final} is:

$$\frac{1}{r_{final}} = \frac{1}{r_{in}} + \frac{1}{r_{th}} \tag{2.14}$$

where the curvature due to the thermal expansion in the two layers ($1 / r_{th}$) is negative (because the heated beam will unfold) and can be expressed as:

$$\frac{1}{r_{th}} = \frac{6 \cdot t \cdot (\alpha_{Cr} - \alpha_{SiO_2}) \cdot (T - T_0)}{\frac{E'_{SiO_2} \cdot t_{SiO_2}^{3}}{E'_{Cr} \cdot t_{Cr}} + \frac{E'_{Cr} \cdot t_{Cr}^{3}}{E'_{SiO_2} \cdot t_{SiO_2}} + 6 \cdot t_{SiO_2} \cdot t_{Cr} + 4 \cdot t_{SiO_2}^{2} + 4 \cdot t_{Cr}^{2}} \tag{2.15}$$

with the defined constant K in section 2.3.2 (Eqn. 2.10), we get

$$\frac{1}{r_{th}} = K \cdot \Delta\varepsilon_{th} \tag{2.16}$$

where $\Delta\varepsilon_{th} = (\alpha_{Cr} - \alpha_{SiO2}) \cdot (T - T_0)$ is the linear strain difference caused by thermal actuation. E'_{SiO2} and E'_{Cr} are the biaxial elastic moduli for the silicon dioxide and the chromium respectively, $t = t_{Cr} + t_{SiO2}$ and t_{SiO2} and t_{Cr} the thicknesses of the two bimorph components. T and T_0 are the working and initial temperature, respectively and α_{Cr} and α_{SiO2} are the coefficients of thermal expansion of the two bimorph layers.

Besides the assumptions listed in section 2.3.2, for simplification, it is supposed that the temperature distribution is uniform over the whole beam and that the Young's modulus, the Poisson's ratio and the coefficient of thermal expansion remain constant with temperature.

Thermal bimorph sensitivity design

The sensitivity for a bimorph actuator is defined as the mechanical deflection angle the cantilever is capable to bend per degree Kelvin. The thermal sensitivity is the product of the thermal curvature (Eqn. 2.15) by the beam length divided by the temperature difference (Eqn. 2.17).

$$s = \frac{L}{r_{th}} \cdot \frac{1}{\Delta T} \tag{2.17}$$

It varies linearly with the difference of the coefficient of thermal expansion of the two component layers. The highest values are achieved by combining SiO_2 with metals such as Au, Cu or Ni. These metals have a reasonably high α (Table 2.4). However, the metals should also be easily to integrate, allowing simple technological steps and MEMS deposition and etching compatibility to the type of processes that have been selected (Chapter 3). Despite the relatively low coefficient of thermal expansion of the chromium, the combination SiO_2 / Cr is finally preferred because of the simple deposition of this metal on the dioxide (no need of seed layer), its excellent adhesion and its compatibility to known fabrication processes (Chapter 3)..

Table 2.4 Thermomechanical properties of bulk materials at room temperature [127, 4, 135].

Material	Young's modulus E [GPa]	Coefficient of thermal expansion α [10^{-6} K $^{-1}$]
Si	162	2.6
SiO_2	75	0.4
Al	69	23
Au	80	14.3
Cr	280	5
Cu	120	16.7
Ni	210	12.8
Pt	147	8.9
Ti	116	8.6

Eqn. 2.12 expresses the relation for a maximal curvature. For given materials and a constant oxide thickness, a maximum curvature can be achieved if the ratio between the thickness of the two layers is inversely proportional to the ratio of the biaxial elastic moduli. In order to obtain maximum actuation sensitivity, i.e.

maximum angular deflection for a given temperature difference (Fig. 2.7), we calculated the chromium thickness. t_{SiO2} is approximately 0.8 µm. For this value, a maximum sensitivity is obtained for $t_{Cr} = 220$ nm. As seen in section 2.3.2, a compromise must be found between high bimorph sensitivity and a reasonable initial curvature position corresponding to the application specifications. Finally, we deposited a $t_{Cr} = 400$ nm because it gives an adequate rest position angle. This results in a sensitivity of about 0.1 °/K.

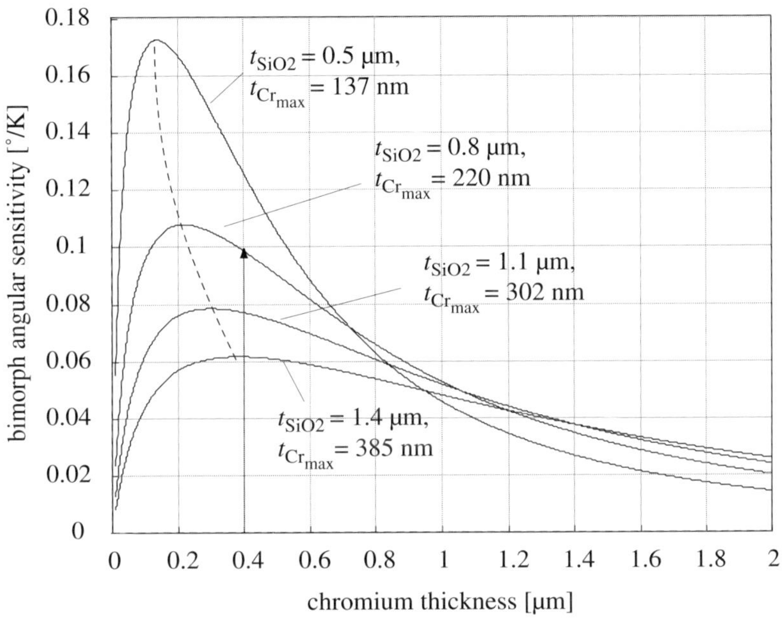

Fig. 2.7 *Calculated actuator sensitivity vs. chromium thickness for various SiO_2 thicknesses, dashed line shows the maximum sensitivity curve. The arrow shows the sensitivity for the current bimorph actuator design corresponding to the material parameters in Table 2.4.*

In order to compare theory and experimental data, the deflection angle has been measured for three devices. The device was heated on a hot plate and the deflection of a reflected laser beam on the mirror was measured on a screen. The measured angle is called the optical angle, which is twice the mechanical angle (effective angle that the mirror moves). Fig. 2.8 plots the optical angle versus beam temperature for a microscanner. Series 1 to 8 were done in chronological order usually beginning with heating up, and cooling down just afterwards. A linear approximation was plotted for each series.

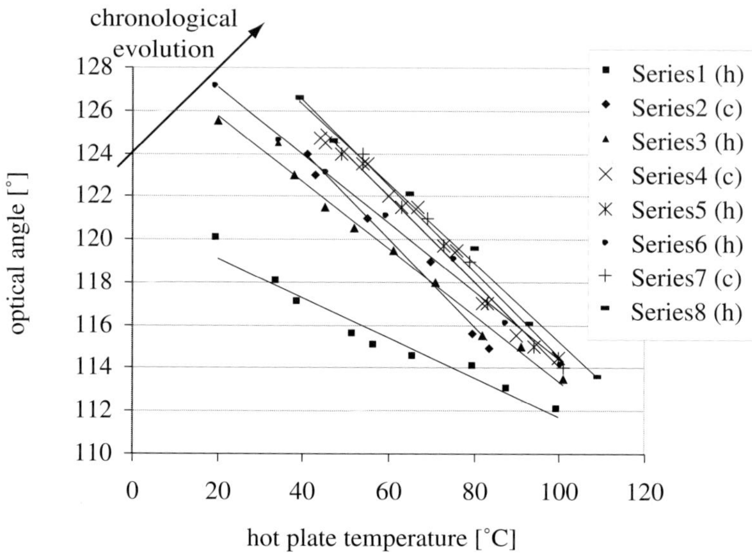

Fig. 2.8 Optical angle vs. temperature for one device. The heating (h) and cooling down (c) is represented chronologically from series 1 to 8. Note that the optical angle is twice the mechanical angle.

One can observe a chronological variation of the slope (thus the sensitivity of the actuator) until it stabilizes (Fig. 2.8). As a matter of fact, this phenomenon was also observed when the actuator is heated with a current. As this feature can be seen even when the microscanner is actuated by external heat, it can be concluded that it is not an intrinsic phenomenon due to the current such as electromigration [125]. It is probably a thermomechanical annealing process, transforming the low quality Cr layer despite the low temperatures (less than 150 °C). Indeed, it is well known that evaporated layers, such as the Cr layer, deposited at low temperatures are porous structures consisting of tapered crystallites [125], resulting in low quality layers. To accelerate this "burn-in" behavior, a careful annealing of the chromium homogenizes the layer and might circumvent, or at least shorten, the "burn-in" phase.

For the last stabilized series, the angular sensitivity can be calculated from the linear approximations. We achieve a measured sensitivity of 0.098 ° / °K for the three devices (from a same wafer), which is comparable to the calculated sensitivity of 0.1 °/°K. The difference between measurement and calculation may come from the measurement inaccuracy which is difficult to estimate because the real temperature of the cantilever was not measured. The coefficient of thermal expansion may be different for thin films [136, 137] as well as the Young's modulus.

Influence of the neglected gold layer (trimorph)

As seen at the beginning of the chapter (Table 2.1), a 30 nm thick gold layer is deposited on top of the Cr layer. In reality, we have a trimorph beam. As gold has a high coefficient of thermal expansion (14.2 10^{-6} K^{-1}), we estimated the influence of this layer. Bühler developed the curvature expression for multimorph cantilevers [4].

$$r = \frac{-\frac{1}{2}\left(\sum_{i=1}^{3} E_i t_i^2\right)^2 + \frac{2}{3}\sum_{i=1}^{3}\sum_{k=1}^{3} E_i E_k t_i^3 t_k + 2\sum_{i=1}^{2}\sum_{k=i+1}^{3} E_i(E_k t_i^2 t_k^2) + 2E_1E_3t_1t_2t_3t_{tot}}{\left(\sum_{i=1}^{2}\sum_{k=i+1}^{3} \alpha_{ki}E_iE_kt_it_k(t_i+t_k) + 2\alpha_{31}E_1E_3t_1t_2t_3\right)\cdot \Delta T} \tag{2.18}$$

with $t_{tot} = t_1 + t_2 + t_3$ and $\alpha_{ki} = \alpha_k - \alpha_i$.

This formula for a trimorph gave a maximum increase of actuator sensitivity of 3.5% (for the given condition of Table 2.2) when the gold layer is considered. The small error allows us to neglect the gold layer. Trimorph actuators will be discussed in more detail in section 7.6. As the characteristics of a three layer stack is often determined by the two thickest layers, we will continue to refer to the actuator principle as thermal bimorph.

Influence of cantilever width

In our actuator, the metal layer (80 µm wide) is narrower than the dioxide layer (100 µm wide). Chu, *et al.* [132] developed the curvature expression considering the width. For the conditions of Table 2.2, the sensitivity has a maximal error of only 2% if the width is neglected.

Conclusions

This section presented the design of the thermal bimorph actuator and the importance of the parameters to consider. A high difference of the coefficients of thermal expansion of the two chosen materials is desired. Technological constraints must also be considered. SiO_2 and Cr are chosen, directed by materials characteristics and especially for their easy integration into the microfabrication process.

For given conditions, a simple relation between thickness ratio and elastic moduli ratio of the two materials maximizes the actuator sensitivity. The optimum design is also a compromise between high actuator sensitivity and adequate initial rest position. A good correlation is found between measured and calculated actuator sensitivity. Both are found to be 0.1 ° / °K.

2.4 STATIC TEMPERATURE DISTRIBUTION IN THE MICROSCANNER

Ideally, the microactuator should have a low power consumption. To reach an optimal thermal efficiency, it is necessary to understand the different effects that can influence the optimum yield of the bimorph actuator. For this, the whole structure must be taken into consideration, which means the actuator attached to the silicon substrate and the mirror at the end of the cantilever actuator, including the surrounding air. Several effects, like temperature distribution in the mirror, thermal losses through the mirror, losses to the substrate and losses to the surrounding gas are of great interest for optimization of the model.

2.4.1 Analytical model

An analytical model for the static temperature distribution in the actuator beam and the mirror has been developed [149]. The basic model takes into account only heat loss by conduction in the beam and the mirror and convection to the surrounding gas, and considers an uniform internal heat source. The radiation can be neglected as we achieve maximum temperature of about 200 °C in the beam. The problem is supposed to be stationary, corresponding to a DC drive of the actuator. Furthermore, as the beam is only about 1.2 µm thick, the temperature can be considered to be uniform throughout the actuator thickness. The mirror is made of copper (6 µm thick). Its thermal conductivity is high, so that the temperature can be considered to be uniform throughout the mirror thickness (z-direction) (Fig. 2.9 a).

The heat transfer in the y-direction is reduced to convection in air. For simplification, we consider the temperature as uniform in the y-direction. Thus, the problem can be simplified to a one-dimensional one, where the temperature T

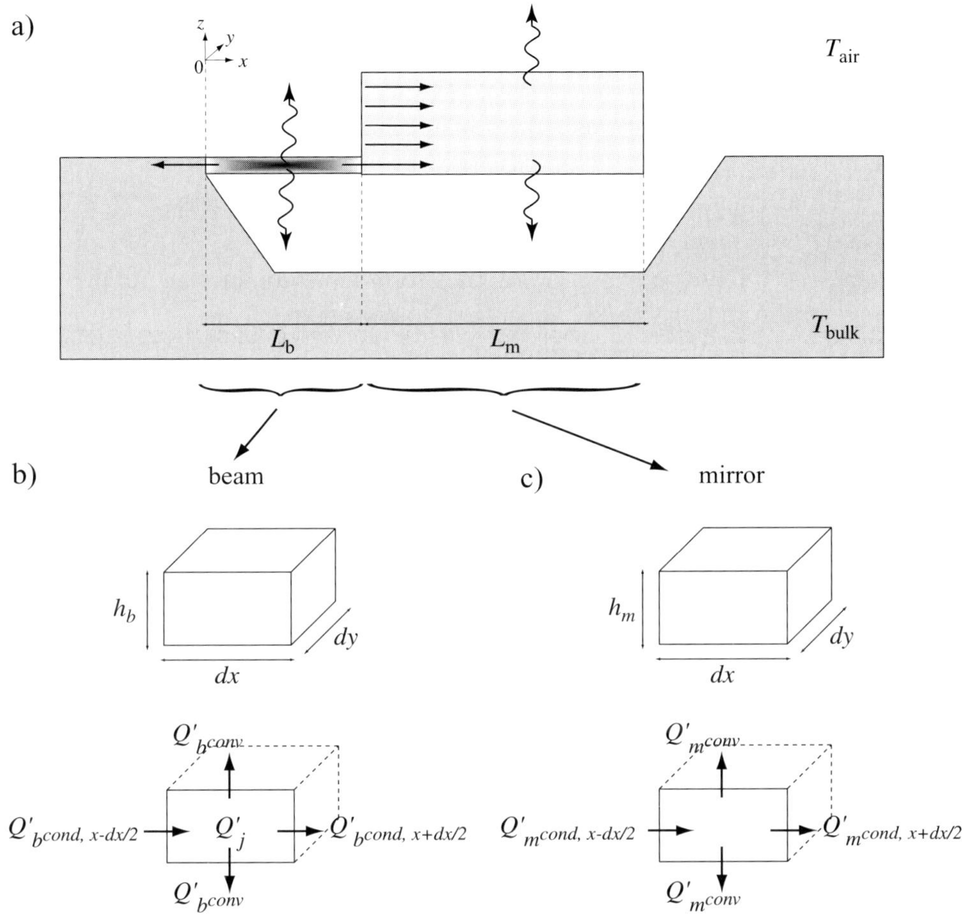

Fig. 2.9 Schematic description of the 1D micromirror model: a) general view; note that for simplification, the device is not curled up; the straight arrows represent the conduction, the curved arrows the convection; b) and c): view of a unit cell and representation of the various heat transfer components on a unit cell for the beam and for the mirror, respectively.

is only a function of the position x along the microscanner (Fig. 2.9 a). In the stationary state, the energy conservation law implies that in the beam,

$$Q'_J - (Q'_{b,\,cond,\,x+dx/2} - Q'_{b,\,cond,\,x-dx/2}) - 2Q'_{b,\,conv} = 0 \qquad (2.19)$$

and in the mirror,

$$Q'_{m,\,cond,\,x+dx/2} - Q'_{m,\,cond,\,x-dx/2} - 2Q'_{m,\,conv} = 0 \tag{2.20}$$

where Q' represents the various heat transfer (in W) from and to the unit cell $h_b \cdot dx \cdot dy$ and $h_m \cdot dx \cdot dy$ (Fig 2.9 b and c). The various indices stand for:

- J internally generated by Joule effect by the resistive heaters,
- b and m for beam and mirror respectively,
- *cond*, $x\text{-}dx/2$ and *cond*, $x\text{+}dx/2$ transferred by conduction in the scanner through the $x\text{-}dx/2$ and $x\text{+}dx/2$ planes respectively,
- *conv* transferred by convection to the air.

The terms of the Eqn. 2.19 and 2.20 can be written as follows:

$$Q'_J = q'_j \cdot dx \cdot dy \tag{2.21}$$

$$Q'_{b,\,cond,\,x+dx/2} = -\lambda_b \cdot \frac{d}{dx}T(x)\Big|_{x+dx/2} \cdot h_b \cdot dy \tag{2.22}$$

$$Q'_{b,\,cond,\,x-dx/2} = -\lambda_b \cdot \frac{d}{dx}T(x)\Big|_{x-dx/2} \cdot h_b \cdot dy \tag{2.23}$$

$$Q'_{b,\,conv} = \alpha_b \cdot [T(x) - T_\infty] \cdot dx \cdot dy \tag{2.24}$$

$$Q'_{m,\,cond,\,x+dx/2} = -\lambda_m \cdot \frac{d}{dx}T(x)\Big|_{x+dx/2} \cdot h_m \cdot dy \tag{2.25}$$

$$Q'_{m,\,cond,\,x-dx/2} = -\lambda_m \cdot \frac{d}{dx}T(x)\Big|_{x-dx/2} \cdot h_m \cdot dy \tag{2.26}$$

$$Q'_{m,\,conv} = \alpha_m \cdot [T(x) - T_\infty] \cdot dx \cdot dy \tag{2.27}$$

where the symbols denote:

- q'_j electric power dissipated by Joule effect per unit surface. (This assumption neglects the temperature dependence of the heating resistance. The positive temperature coefficient of metals increases the electrical resistance at the hot tip of the beam. This increases the voltage drop and the heating power. To keep the model simple, this effect will be neglected here.)
- h_b and h_m beam and membrane thickness, respectively,
- λ_b and λ_m thermal conductivity of beam and of mirror,
- α_b, α_m average convection heat transfer coefficients in the air for the beam and the mirror respectively,

- T_∞ temperature of the air far from the heating cantilever.

Let us assume that T_∞ is the same on both sides of the mirror or the beam and that it is equal to the air temperature T_{air}. By inserting Eqn. 2.21 to 2.24 into Eqn. 2.19 and Eqn. 2.25 to 2.27 into Eqn. 2.20, we obtain

$$q'_j \cdot dx \cdot dy + \lambda_b \cdot \frac{d^2}{dx^2}T(x) \cdot dx \cdot h_b \cdot dy \tag{2.28}$$
$$-(2\alpha_b \cdot [T(x) - T_{air}] \cdot dx \cdot dy) = 0$$

$$\lambda_m \cdot \frac{d^2}{dx^2}T(x) \cdot dx \cdot h_m \cdot dy \tag{2.29}$$
$$-(2\alpha_m \cdot [T(x) - T_{air}] \cdot dx \cdot dy) = 0$$

which can be simplified to:

$$q'_j + \lambda_b \cdot h_b \cdot T''(x) - 2\alpha \cdot [T(x) - T_{air}] = 0 \tag{2.30}$$

$$\lambda_m \cdot h_m \cdot T''(x) - 2\alpha \cdot [T(x) - T_{air}] = 0 \tag{2.31}$$

We replaced the expressions α_b and α_m by α, where α is the general average convection coefficient for the microscanner. Finally, the following system of second order linear differential equation is obtained:

$$T''(x) - \frac{2\alpha}{\lambda_b \cdot h_b}T(x) = -\frac{q'_j + 2\alpha T_{air}}{\lambda_b \cdot h_b} \tag{2.32}$$

$$T''(x) - \frac{2\alpha}{\lambda_m \cdot h_m}T(x) = -\frac{2\alpha T_{air}}{\lambda_m \cdot h_m} \tag{2.33}$$

The solution of the differential Eqn. 2.32 is the following:

$$T_b(x) = C_1 \cdot e^{\sqrt{\frac{2\alpha}{\lambda_b \cdot h_b}} \cdot x} + C_2 \cdot e^{-\sqrt{\frac{2\alpha}{\lambda_b \cdot h_b}} \cdot x} + C_3 \tag{2.34}$$

which is **the static temperature distribution in the beam** and for Eqn. 2.33:

$$T_m(x) = C_4 \cdot e^{\sqrt{\frac{2\alpha}{\lambda_m \cdot h_m}} \cdot x} + C_5 \cdot e^{-\sqrt{\frac{2\alpha}{\lambda_m \cdot h_m}} \cdot x} + C_6 \tag{2.35}$$

which is **the static temperature distribution in the mirror**, where $C_{1,2,3,4,5,6}$ are constants. By inserting the expressions of $T(x)$ in the differential equations, it is found that

$$C_3 = \frac{q'_j}{2\alpha} + T_{air} \tag{2.36}$$

$$C_6 = T_{air} \tag{2.37}$$

The four other constants are determined from the boundary conditions, which are the following:

- At $x = 0$ (Fig. 2.9 a), the temperature in the beam is the same as in the Si substrate, thus Eqn. 2.34 can be written as

$$T_b(0) = T_{bulk} \tag{2.38}$$

- At $x = L_b$, the temperature in the beam is the same as in the mirror, so

$$T_b(L_b) = T_m(L_b) \tag{2.39}$$

- At $x = L_b$, the heat transfer rate by conduction is the same in the beam as in the mirror, therefore:

$$-\lambda_b \cdot \frac{d}{dx} T_b(x)\big|_{L_b} \cdot h_b \cdot b_b = -\lambda_m \cdot \frac{d}{dx} T_m(x)\big|_{L_b} \cdot h_m \cdot b_m \tag{2.40}$$

where b_b and b_m are the width of the beam and the mirror, respectively. Note that, for simplification, the conduction is considered to be uniform along the mirror thickness, even if the latter is six times larger than beam thickness.

- At $x = L_b + L_m$, at the end of the mirror, the heat transfer rate by conduction is equal to zero.

$$\frac{d}{dx} T_m(x)\big|_{L_b + L_m} = 0 \tag{2.41}$$

The system of Eqn. 2.38 to 2.41 was solved in a Matlab program. As the expressions of $C_{1,2,3,4}$ are lengthy and complicated, they are not expressed here (Appendix A.2).

The expression of the temperature distribution along the direction x is given by Eqn. 2.34 for the beam and by Eqn. 2.35 for the mirror.

2.4.2 Determination of constants

The basic model takes into account only heat loss by conduction in the beam and the mirror and convection to the surrounding gas. It considers an uniform internal heat source of 0.167 W/mm^2, corresponding to an applied power of 5 mW on the beam. We should remind that the beam is made of silicon dioxide (800 nm thick), chromium (370 nm thick) and gold (30 nm thick). For the Au/Cr/SiO_2 beam a mean thermal conduction λ_b for the beam is defined considering the contribution of each layer (Table 2.7). The mirror is made of copper (6 μm thick) and Au/Cr (maximal 400 nm thick), as reflecting layer. The latter is neglectable relatively to the copper layer. Thus, the mirror conductivity is equal to thermal conductivity of the copper layer (Table 2.5).

Table 2.5 Constants used for calculating T(x) with Eqn. 2.34 and 2.35.

Constant	Value
q'_j	0.167 W/mm^2
α_{free}	2 .. 25 $W/m^2 \cdot K$
α_{forced}	25 .. 250 $W/m^2 \cdot K$
λ_b	30 W/m·K
$\lambda_m = \lambda_{Cu}$	401 W/m·K
h_b	1.2 μm
h_m	6.4 μm
L_b	300 μm
L_m	500 μm
b_b	100 μm
b_m	300 μm
T_{bulk}	293 K
T_{air}	293 K

Free and forced convection

The study above is done for a stationary state, thus the temperature is stable in time and so does the position of the device. It is considered that the microdevice is in a free convection state when driven in a DC mode.

If actuated at the resonance frequency (AC mode) and deflecting from one position to the other, the mirror and the beam are subjected to a forced convection. Furthermore, at resonance, which occurs above the thermal cut-off of the bimorph actuator, the total temperature variation in time is low. It is considered to be also in a stationary state. In this case, the Reynolds number (*Re*) was calculated (Eqn. 2.42) to estimate the nature of the flow when the mirror is deflecting.

$$Re = \frac{u \cdot D}{\upsilon} \tag{2.42}$$

where u is the gas velocity, D the diameter or the height (500 µm) of the mirror and υ the kinematic viscosity of the gas ($16 \cdot 10^{-6}$ m^2/s for air at 300 K). In our case, the average driving frequency at resonance is about 400 Hz for a 1D scanner. The distance from substrate to middle of the mirror is about 500 µm and the maximal mechanical angle performed at resonance is over 90°. Thus, the velocity is estimated at 0.22 m/s, which leads to a Reynolds number of about 10. For a vertical flat plate, with a finite length, the system at a $Re = 10$ corresponds to the transition zone between a laminar and a turbulent flow [148, 149]. Thus, at resonance, the device is exposed to a transition flow.

The average convection coefficient

The effective value of α is very difficult to determine accurately and it depends on the type of drive mode, thus the convection type. We will separate the two types.

- *Free convection*

It has been seen above that in a DC mode, we are in a free convection system and that the range of commonly found values for free convection is between 2 and 25 W/m^2·K [149].

Fig. 2.10 represents the temperature distribution as a function of the length of the microscanner (x) calculated with the constants given in Table 2.5. The calculation was done for the beam without mirror, in the case of no convection ($\alpha = 0$) and for $\alpha = 10$ W/m^2·K. One can see that convection in the beam is not determinant in a free convection mode. This is due to the fact that only a small part of the heat is brought by convection to the air (for the beam), and that the

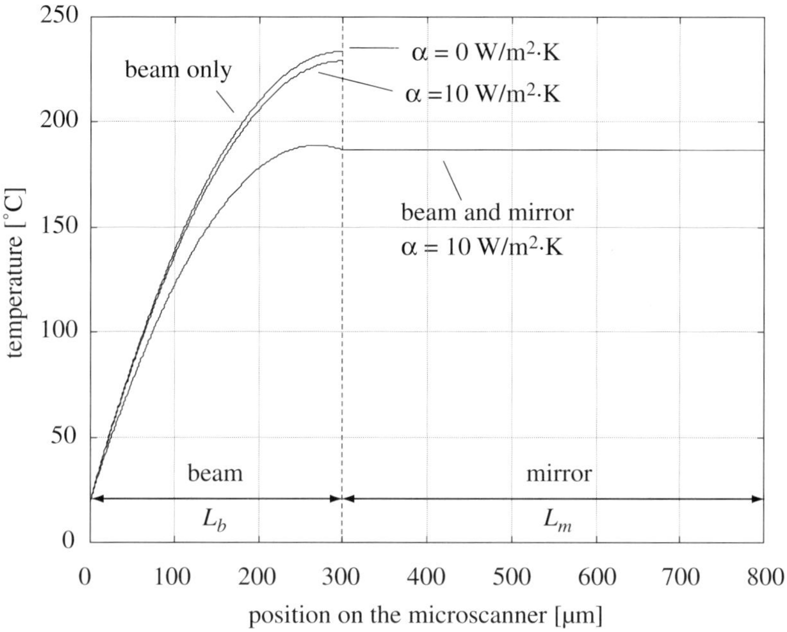

Fig. 2.10 Temperature profile versus position on the microscanner (in x-direction) for the constants given in Table 2.5 and in the case of free convection.

largest part of the generated heat is lost by conduction in the bulk silicon. As a matter of fact, a heat transfer rate by conduction (proportional to the curves slope at $x = 0$) in the substrate of about 4.9 mW is calculated (for $\alpha = 10$ W/m^2·K). As the total applied power is 5 mW, the heat lost by convection is about 0.1 mW, thus 2% of the injected power.

The last curve represented in Fig. 2.10 corresponds to the temperature distribution in the whole microscanner for $\alpha = 10$ W/m^2·K. It can be calculated that almost 10% (0.49 mW) of the generated heat is lost by conduction in the mirror and 88% (4.4 mW) is lost in the silicon substrate.

For the following analysis, we assume an arbitrary α of about 10 W/m^2·K in the free convection mode.

- *Forced convection*

At resonance, the device is in a forced convection mode and exposed to a transition flow. Moreover, in this case, it is really difficult to estimate experimentally the temperature in the beam. For a forced convection mode, the range of commonly found values is between 25 and 250 W/m^2·K [149]. Fig. 2.11

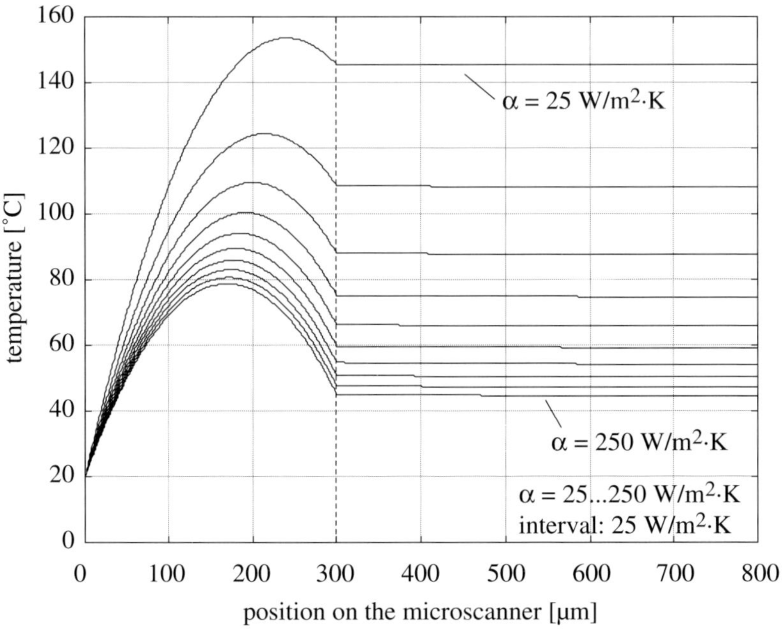

Fig. 2.11 Temperature profile versus position on the microscanner (in x-direction) for the constants given in Table 2.5 and in the case of forced convection, α is varying.

plots the temperature distribution for regularly spaced values of α between 25 and 250 $W/m^2 \cdot K$.

For $\alpha = 200$ $W/m^2 \cdot K$, it can be calculated that almost 36% (1.8 mW) of the generated heat is lost by conduction in the mirror and 52% (2.6 mW) is lost in the silicon substrate. The heat lost by convection in the beam can be deduced and is about 0.6 mW, thus 12% of the injected power.

As the mirror is at about 51 °C, the heat transfer rate by convection in the mirror has a value of about 6200 W/m^2 in the case of $\alpha = 200$ $W/m^2 \cdot K$. Thus, a power of about 0.93 mW is lost per mirror face, or 1.86 mW on the whole plate. We just saw above that 1.8 mW is lost by conduction from the beam to the mirror. This confirms the Eqn. 2.31, where the absolute value of losses per conduction in the mirror equals the absolute value of losses per convection.

For the following analysis, we assume an arbitrary α of about 200 $W/m^2 \cdot K$ in a forced convection mode. Summarizing, the next table indicates the corresponding convection coefficient used for the following analysis.

Table 2.6 Average convection coefficient used for the different driving modes.

Mode	Constant	Value [W/m^2·K]
free convection, static operation	α_{free}	10
forced convection, resonance drive	α_{forced}	200

Convection versus designed beam length

For long beams convection becomes more important over conduction and the temperature distribution flattens out over the beam length. The ratio between both cooling mechanisms can be calculated regarding the total cooling power by conduction into the substrate (neglecting the mirror) given by

$$P_{cond} = \lambda_b b_b h_b \cdot \frac{d}{dx}T(0) = \frac{-q_j' b_b}{a} \cdot \left(\frac{1}{1+e^{2aL_b}} + \frac{1}{1+e^{-2aL_b}}\right) \tag{2.43}$$

$$a = \sqrt{\frac{2\alpha}{\lambda_b h_b}}$$

compared to the total heating power $P_{tot} = q_j' b_b L_b$. Fig. 2.12 shows how the relative contribution of cooling by convection grows with the length of the cantilever.

2.4.3 Influence of the different constants on the temperature distribution

Influence of beam heat conduction

For the case of a static drive of the micromirror, it is noted that the influence of the heat conduction in the beam is important (Fig. 2.13) for a given thermal conduction in the mirror (Cu, in our case). It is clear that for a small heat loss in the beam, less power is needed to achieve a certain temperature. There are not many conductive materials with lower thermal conductivity (Table 2.7) than our bimorph beam, which is already very low (30 W/m·K). The same remarks as the ones made for the static operation are true for resonance actuation (Fig. 2.14).

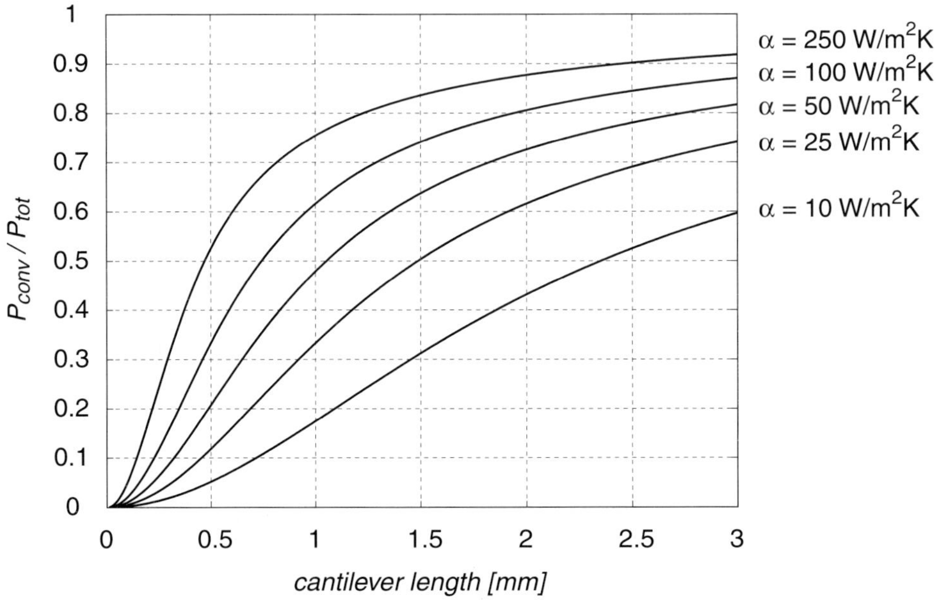

Fig. 2.12 The relative contribution of cooling power by convection grows with the total length of the cantilever (without mirror at the tip).

Influence of convection

As reported in section 2.4.2, convection is not neglectable in our system, independently from the operation mode. It is clear that its influence is increased drastically in a forced convection mode where it has been seen that the convection losses by mirror surface is important (about 1.8 mW).

Influence of geometrical dimensions

When the beam length is increasing, the average temperature increases in the actuator for a given power in any driving mode (Fig. 2.15). Thus, the losses by conduction in the surrounding material are less important when the beam is longer. Indeed, this has been observed experimentally: shorter beams show a higher power consumption than longer beams. At resonance, to achieve an optical scan angle of 80°, an average power of 2.5 mW is needed for a beam length of 300 μm. For the same scan angle, a 120 μm long beam requires 7 mW.

The length of the beam can not be increased indefinitely as we are limited by the initial position of the mirror (due to the residual stress) and the sensitivity to vibrations and shocks as well as mechanical resonance behaviour. A compromise has to be found between thermal efficiency and mechanical stability.

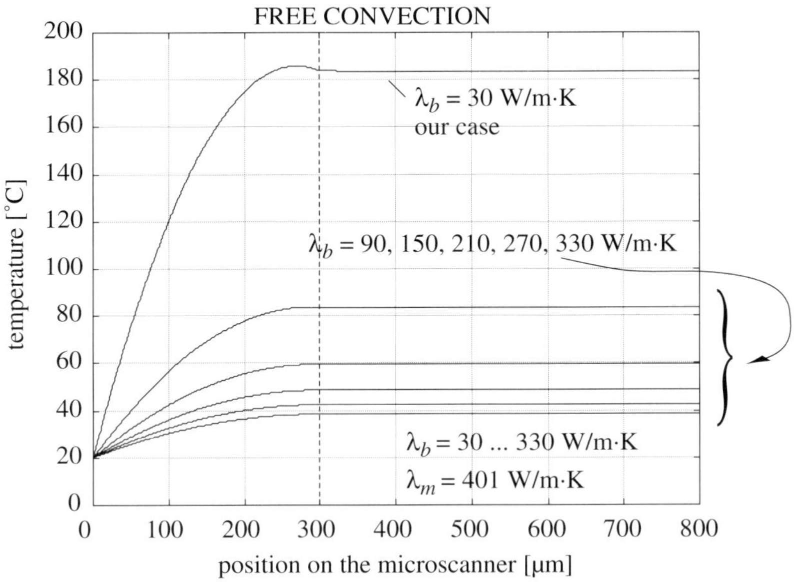

Fig. 2.13 Temperature profile versus position on the microscanner for the constants given in Table 2.5 for α_{free} = 10 W/m^2·K, λ_b is varying.

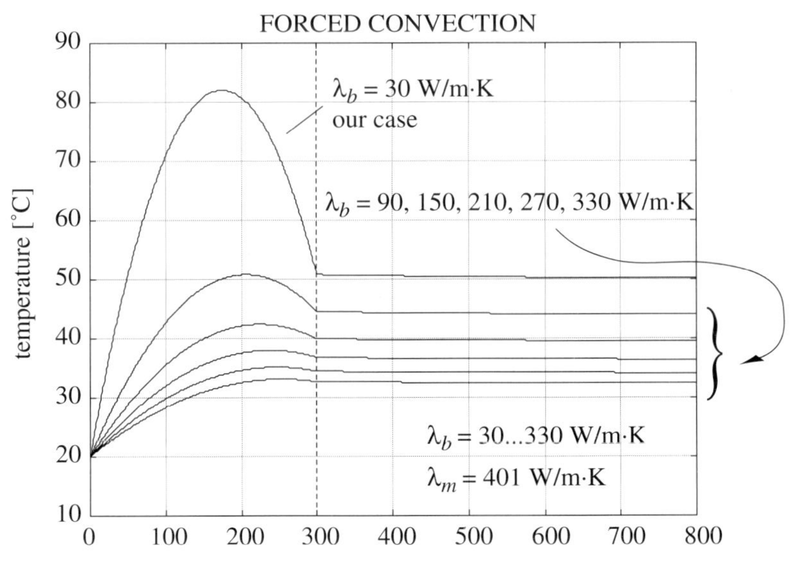

Fig. 2.14 Temperature profile versus position on the microscanner for the constants given in Table 2.5 for α_{forced} = 200 W/m^2·K, λ_b is varying.

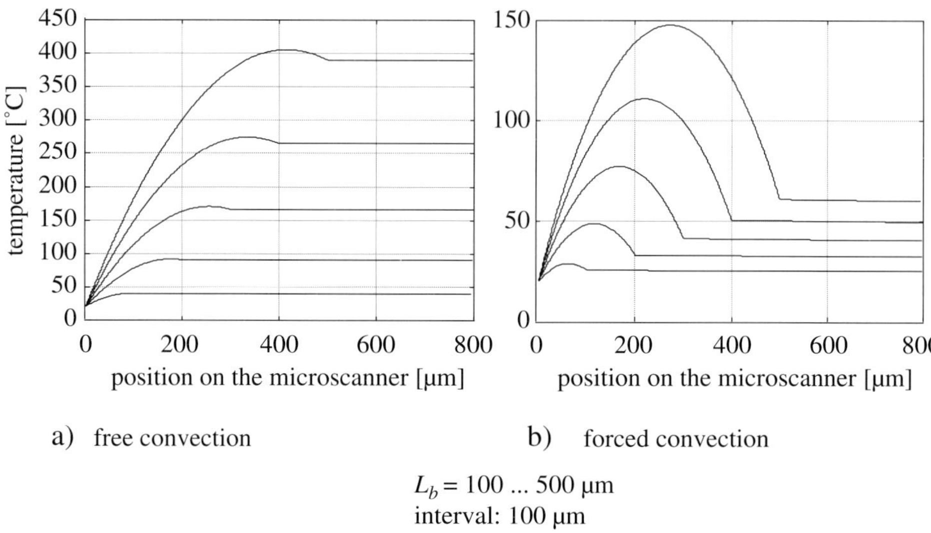

Fig. 2.15 Temperature profile versus position on the microscanner (in x-direction) for the constants given in Table 2.5 and 2.6.
The length of the beam is changing a) in the case of free convection; b) in the case of forced convection.

The mirror area has major influence on the heat transfer of the microscanner in the dynamic mode, due to the heat losses by convection. So does the beam area. The influence of these parameters during static operation is quite low. Thus, large areas are suitable in a dynamic operation of the transducer. They are also suitable for optical reasons (diffraction) but not always for the mechanical resonance behaviour (large mirrors tend to low resonance frequencies) and for fabrication reasons (sticking problems).

When the thickness of the beam decreases, its average temperature increases for the same applied power.

2.4.4 Summary

In free convection mode, the losses by convection are not determinant in the beam (with or without mirror) and represent 2% of the generated heat in the case of a convection coefficient of $\alpha = 10$ W/m^2·K. The largest part of the generated heat is lost by conduction in the silicon substrate (88%) and in the mirror (10%). The influence of heat conduction in the beam is an issue in the microscanner temperature distribution. With a bimorph beam Au/Cr/SiO_2, the thermal conductivity is low and we already have adequate properties.

Convection losses through the mirror are not neglectable in the microscanner, independently from the operation mode. It has been calculated that 10% of the generated heat in a free convection mode ($\alpha = 10\ \mathrm{W/m^2{\cdot}K}$) and 36% of the generated heat in a forced convection ($\alpha = 200\ \mathrm{W/m^2{\cdot}K}$) is lost by conduction in the mirror, thus lost by convection to the surrounding air.

For a given applied power on the beam, the average temperature increases as the beam length increases and the beam thickness decreases. A compromise has to be found between thermal efficiency and mechanical stability. For efficient heat exchange, large mirror areas are necessary, especially in a resonance actuation.

2.5 Response time of the bimorph beam

2.5.1 Thermal cut-off

The effective response time of the bimorph beam is of great interest, in particular for a resonance driving mode. The cut-off frequency is the frequency at which the gain of a transducer drops by 3 decibels below the low-frequency gain value [151]. For higher frequencies, the amplitude of the scanner is frequency dependent. Knowing the *Q* factor (quality factor) of the scanner and the resonance frequency (Chapter 4), the scan amplitude can be deduced. This can be useful for design issue applicated to specific applications.

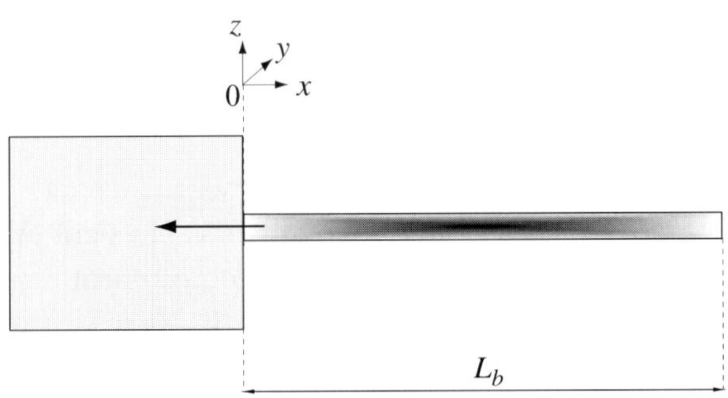

Fig. 2.16 Schematic of the clamped bimorph beam.

The factors affecting the cut-off frequency are calculated in the following. For simplification, the convection is not taken into account. The heat transport equation expressed for a clamped beam in a simplified 1D model (x-direction, Fig. 2.16) can be expressed as [149]:

$$\rho_b \cdot c_b \cdot \frac{\partial T}{\partial t} = \lambda_b \cdot \frac{\partial^2 T}{\partial x^2} + P_{el} \tag{2.44}$$

where P_{el} denotes the mean electric power per volume dissipated in the heating resistor, λ_b the mean thermal conductivity, ρ_b the mean mass density, c_b the mean specific heat capacity of the beam and t the time. At $T(t) = T_{max}$ (reached at beam extremity in a stationary state, $x = L_b$), Eqn. 2.42 can be expressed as :

$$\underbrace{\frac{\rho_b \cdot c_b \cdot L_b^2}{2\lambda_b}}_{\tau} \cdot \frac{\partial T_{max}}{\partial t} - T_{max} = 0 \tag{2.45}$$

where τ is the response time (time constant) of the cantilever. It is the characteristic time it takes for the cantilever to come to rest to a new position, after the quantity of power applied has been abruptly changed [151].

A thermal diffusivity constant D [149, 152] can be defined (Eqn. 2.46):

$$D = \frac{\lambda_b}{\rho_b \cdot c_b} \tag{2.46}$$

and the time constant τ of the bimorph beam can be expressed as [153]:

$$\tau = \frac{L_b^2}{2D} \tag{2.47}$$

The cut-off frequency depends on the thermomechanical properties of the layers and goes with the $1 / L_b^2$. Table 2.7 gives the physical properties of some material of interest, their corresponding volumetric heat capacity and their thermal diffusivity constant D. The latter gives the speed to respond to a heat excitation. In our case, an equivalent $\lambda_b = 30$ W/K·m and equivalent volumetric heat capacity of $\rho_c = 2.53 \cdot 10^6$ J/m^3·K can be calculated for the three layers (Au/Cr/SiO_2), thus a D of $11.83 \cdot 10^{-6}$ m^2/s. This results in a thermal constant time of 3.8 ms for a 300 µm long beam.

2.5.2 Measurements of the cut-off frequency of the bimorph actuator

Fig. 2.17 plots the measured deflection of a simple beam (without mirror) versus driving frequency (an AC sinusoidal voltage is applied and the deviation

Table 2.7 Thermomechanical properties of bulk materials at room temperature [127, 4, 135].

Material	Thermal conductivity λ [W/K·m]	Specific heat capacity c [J/g·K]	Mass density ρ [10^6·g/m^3]	Volumetric heat capacity $\rho \cdot c$ [10^6·J/m^3·K]	Thermal diffusivity constant D [10^{-6}·m^2/s]
Si	170	0.691	2.42	1.67	101.7
SiO_2	1.3	0.84	2.66	2.23	0.6
Al	235	0.879	2.6	2.28	102.8
Au	318	0.129	19.4	2.5	127
Cr	69	0.449	7.1	3.19	21.6
Cu	401	0.387	8.95	3.46	115.8
Ni	91	0.444	9.04	4.01	22.7
Pt	73	0.133	21.5	1.47	25.5
Ti	15.5	0.47	4.5	2.86	7.3

of the reflected laser on the beam is projected on the position sensitive detector). One can deduce graphically a thermal cut-off frequency of about 150 Hz, thus a response time of 6.7 ms. The differences with the calculated value (3.8 ms) can be explained by inaccuracies on the thermomechanical material properties (the thermal conductivity of the Cr is probably lower than the value given by the tables, due to the layer quality).

2.5.3 Conclusion

The previous calculation is only coarse because the temperature profile changes dynamically, but it gives the design rules that the thermal cut-off frequency goes down with the square of the beam length. Fast response time requires high thermal diffusivity constants, thus high thermal conductivities and short beams. This is in contradiction with the requirements for a low power consumption beam (section 2.4). Depending on the application, a compromise has to be found between low power consumption and short response time.

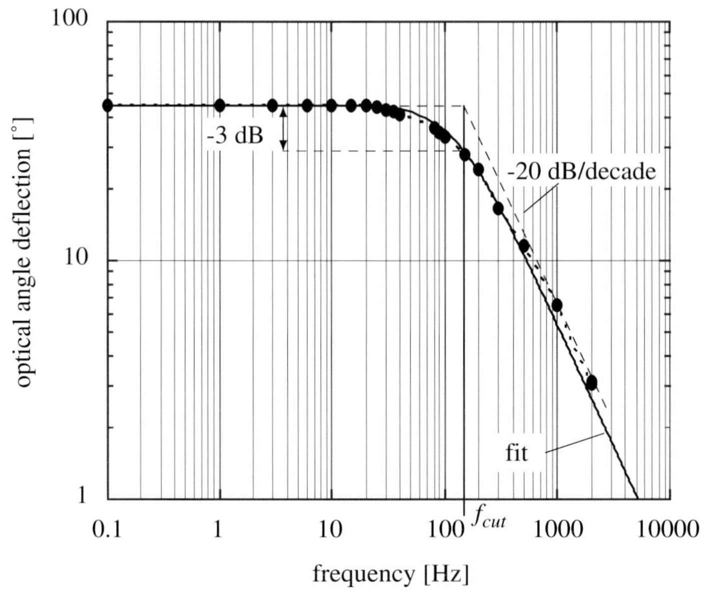

Fig. 2.17 Measured frequency response curve for a 300 μm long bimorph beam (dots) [154] and corresponding fit of a first order harmonic system (full line).

2.5.4 Dynamic temperature distribution in the bimorph beam

In order to get closer to the real conditions, namely to a dynamic temperature profile in the beam, a simulation model was adapted for straight beams attached at one end (Fig. 2.16) from a 1-D model developed by Weber [155]. It is a simple numerical method which was be easily implemented on a Matlab program. Losses are due to radiation, convection and conduction through the extremity. The thermal behavior of the micro-beam heated by a pulsed current is given by Eqn. 2.48:

$$\rho_b \cdot c_b \cdot \frac{\partial T}{\partial t} = \lambda_b \cdot \frac{\partial^2 T}{\partial x^2} + P_{el} + \frac{-\alpha S(T - T_{air}) - \varepsilon \sigma S T^4}{V} \tag{2.48}$$

where besides the constants already described for Eqn. 2.44, ε represents the emissivity, σ the Stefan-Boltzmann constant, α the convection coefficient, S the surface and V the volume of the beam.

The solving method for the non-linear second order differential Eqn. 2.48 is a finite scheme based on the Euler approximation (for more details, see [155]). At time $t = 0$, the entire micro-beam is at room temperature and the extremity of the filament is fixed on the bulk silicon substrate. Eqn. 2.48 was solved considering the constants of Table 2.5. Moreover, the emissivity ε is considered to be equal to zero. The two following graphs represent the dynamic temperature profile as a function of time of the solved Eqn. 2.48.

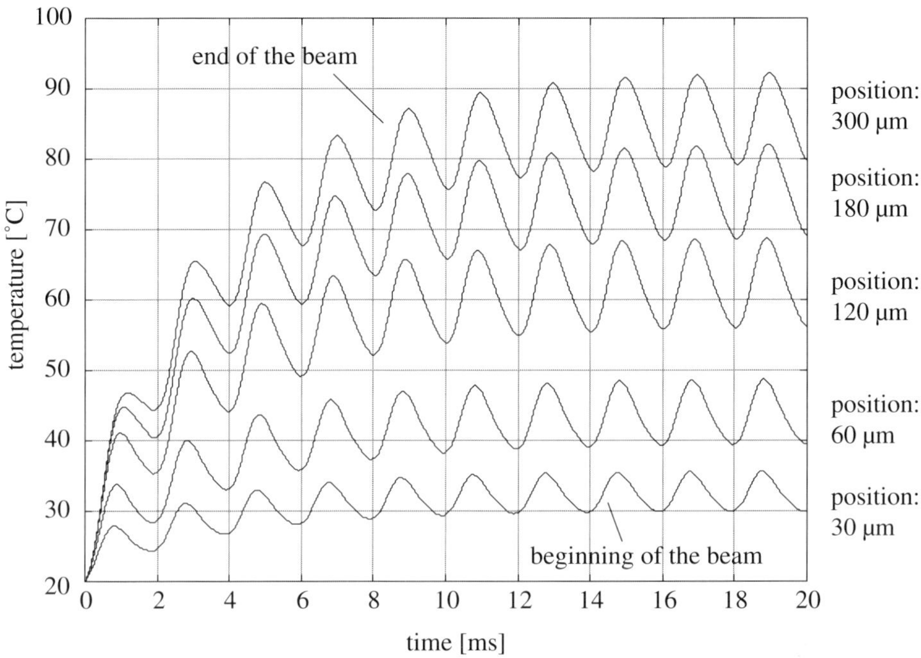

Fig. 2.18 Dynamic temperature profile (for sinusoidal excitation) versus time for different positions along the beam (total length of 300 μm) for the values of Table 2.5 and a convection coefficient $\alpha = 10\ W/m^2 \cdot K$. *A sinusoidal current at 500 Hz was applied.*

From Fig. 2.19 one can observe that from the middle of the beam on to the end of the cantilever, there is no significant change in the dynamic variation of temperature; the amplitude remains the same. A pulsed sine signal at 500 Hz was applied. At the beginning of the beam (position: 30 μm), the amplitude of the temperature is lower than in the rest of the cantilever. This is due to the conduction losses in the substrate.

The convection has no influence on the dynamic temperature variation, as shown in Fig. 2.19. However, it changes the average temperature in the cantilever. When the convection gets important ($\alpha = 200\ W/m^2 \cdot K$), the maximum temperature is reached faster than for lower convection coefficient.

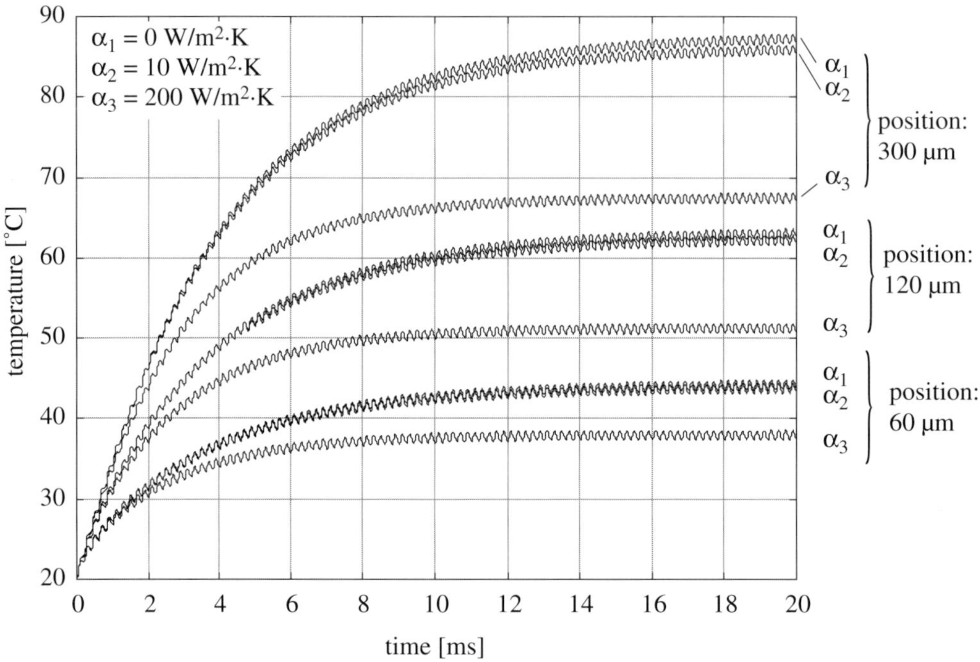

Fig. 2.19 Dynamic temperature profile versus time at different positions along the beam (total length of 300 µm) and for various convection coefficients (values of Table 2.5). A sinusoidal current at 5000 Hz was applied.

From these calculations, a thermal cut-off frequency for a 300 µm long beam of 150 Hz was found, which is a good correlation with the measured value (also about 150 Hz, section 2.5.2). This value is about 35 Hz when the cantilever becomes 600 µm long, confirming the reciprocal dependency of the thermal cut-off frequency to the square of the beam length .

2.6 General conclusions

The purpose of this chapter was to study and understand the thermal bimorph actuator of a microscanner device in order to optimize the actuator sensitivity and to meet performance specifications for given applications.

The determination of the initial position of the mirror is directly dependent to the internal stress in the layers. The reproducibility of this stress is fairly well controlled on a same wafer batch. Despite the fact that this preliminary position will change when the microscanner is operated for the first time, the final "burned-in" position is stable and depends on the very first angle. Bimorph materials and actuator dimensions were designed for optimum actuator properties, like initial position, sensitivity and speed. A compromise had to be found between materials with interesting characteristics for bimorph sensitivity and materials adapted to the chosen technological process.

The calculated values of initial position and sensitivity showed good correlation with the measurements. A thermal analysis was done for the microscanner. In a free convection mode for the bimorph beam (with or without mirror) the largest part of the generated heat is lost by conduction in the silicon substrate (88%) and in the mirror (10%). Convection cooling by the mirror is not neglectable, independently from the operation mode. In particular, it has been calculated that 36% of the generated heat in forced convection actuation (for the chosen convection coefficient) is lost by conduction into the mirror, thus by convection to the surrounding air.

For a given applied power on the beam, the average temperature increases as the beam length increases and the beam thickness decreases. A compromise was found between thermal efficiency and mechanical stability. The length of the beam influences also the actuator response time. A trade-off has to be found between thermal efficiency and fast response time. For efficient heat exchange, large mirror areas are necessary, especially in resonance actuation.

The characterization of the actuator showed that an initial thermo-mechanical "burn-in" phase is necessary to stabilize the actuator sensitivity. The mechanical and electrical characteristics of the conductive layer, depending on the layer morphology, are directly involved in actuator reliability.

The mentioned microactuator specifications in section 2.1 are met well. Standard materials were chosen, adapted to known fabrication processes. In a static operation mode, the relatively low power consumption (10 mW) for large optical angles of 30° is a satisfying result. We need about 25° mechanical angles for barcode reading applications. As we planned to use the 1D scanner at resonance frequency, we can already expect larger angles at lower power consumption.

3 MICROSCANNER TECHNOLOGY

The microscanners are based on a simple low-cost process, using standard materials and techniques. This chapter presents the fabrication technologies using standard silicon micromachining [107, 71] and process improvements to achieve acceptable yield and reliable devices for reasonable manufacturing costs.

3.1 FABRICATION PROCESS

The devices are obtained by a three-mask, single-side silicon micromachining process. Starting with 1 μm oxidized silicon wafers (<100> crystal orientation), openings are patterned into the oxide by photolithography and BHF etch. These openings define the beams and the zone where the mirror will be deposited and subsequently released. A PVD (evaporation) aluminum sacrificial layer (100 nm) is deposited and structured by lift-off (Fig. 3.1 a-c). When the microdevice is released, the Al layer will allow the mirror to be detached before the beams, preventing the latter to collapse.

A 30 / 370 nm thick PVD (evaporation) Au/Cr layer is then deposited and patterned, defining the conductors and mirror plating base (Fig. 3.1 d-e). The mirrors are then grown by copper electroplating (Fig. 3.1 f). In order to achieve flat mirrors, copper thicknesses of typically 6 μm are deposited. A standard thick positive photoresist for electroplating is used. To eventually use both sides of the mirror (not only the Cr side) and also to homogenize the Cu thickness, the mirror is polished with diamond suspension solutions. This reduces the inhomogeneity close to the average roughness of an evaporated metal layer on polished monocrystaline silicon. The resulting structure is illustrated in Fig. 3.1 g (device top view) and h (cross-section of device before release).

Before releasing the microscanners, the chips of 2.5 by 2.5 mm^2 are presawed to about 2/3 of the total wafer thickness. This keeps the mirrors optically clean avoiding silicon dust contamination later. Additionally, one can continue to process complete wafers.

The future moving part is now released in one single etching step in KOH, which attacks first the Al sacrificial layer liberating the large mirror surface and then the silicon, underetching the beams. The latter are at an angle of 45° with

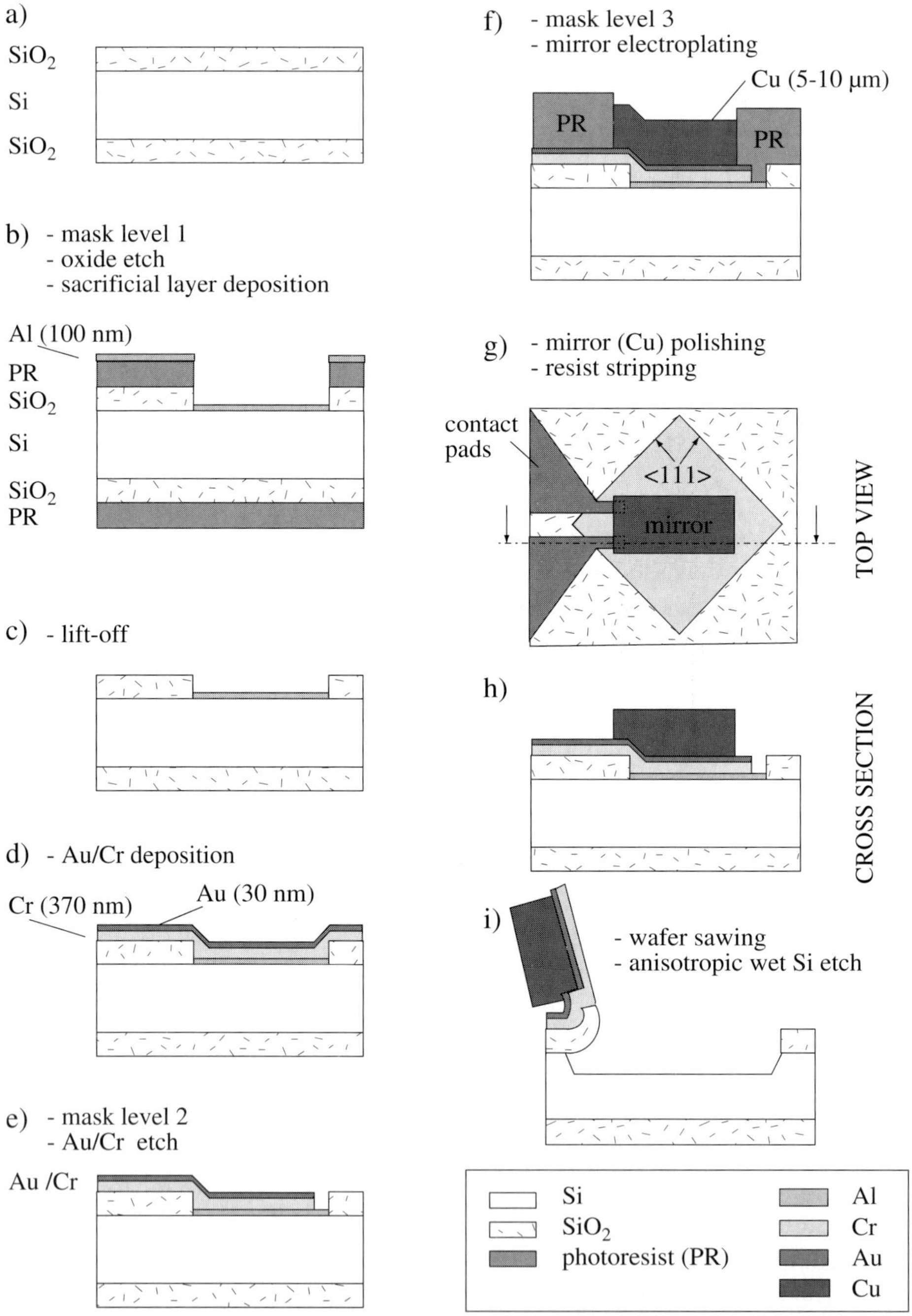

Fig. 3.1 Basic process steps: example of a 1D micromirror.

regard to the <111> crystalline direction of the silicon substrate (Fig. 3.1 g). The required etch depth for release of the structures corresponds to about half the width of the beams. In our simple-beam configuration, the beams are 100 μm wide, thus a 50 μm deep etch is needed, as the beams are underetched from both sides.

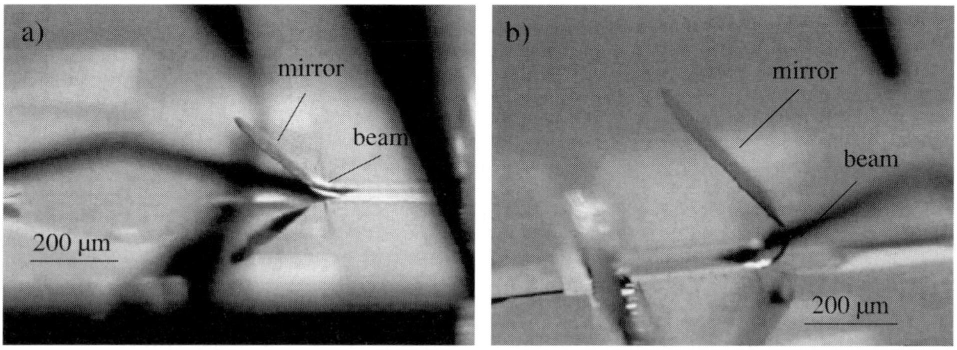

Fig. 3.2 Released 1D microscanner pictures: double beam configuration.
a) Oval mirror (300 μm by 500 μm) and 120 μm long beam.
b) Rectangular mirror (500 by 800 μm^2) and 300 μm long beam.

The released beams curl up, lifting the mirror and bending it backwards (Fig. 3.1 i) by 130° in the case of a 300 μm long beam (Fig. 3.2 b). If the beam is only 120 μm long, it is lifted up to 40° (Fig. 3.2 a).

The back side of the mirror can be used as a reflector. It features an excellent surface quality, as it is formed by a thin film directly evaporated on monocrystaline silicon. The rest position of the released devices, as seen in Chapter 2, is very reproducible within a few degrees on one wafer, due to the homogeneity of the two bimorph components.

Drying the structures after the wet etch is the delicate step. High surface tensions of the rinse liquid come into play during drying [156, 157]. These tensions tend to collapse the structures or cause sticking between the microscanner and the substrate. We achieved a yield of about 10% by carefully dipping the structures into different rinse liquids and solvents, and then leaving them to dry in air. A freeze-drying method using 2-methyl-2-propanol (also known as t-butyl alcohol) was tested by Takeshima *et al.* [158] for MEMS fabrication. This method was developed for biological specimens for scanning electron microscopy [159]. As the 2-methyl-2-propanol is solid at room temperature (the freezing point is 25.5 °C), it is possible to perform freeze-drying without special cooling equipment, at atmospheric pressure. It requires a hot plate, a standard refrigerator and a simple vacuum system for vapour extraction.

After KOH etching and rinsing with water and melted 2-methyl-2-propanol, the microstructures are refrigerated and the rinsing solution that covers the microscanners is frozen. Then, at atmospheric pressure, the frozen solution is sublimated during 12 hours. Freeze drying with 2-methyl-2-propanol instead of water has the advantage that it creates less stress than water freezing. With this technique, the fabrication yield could be raised up to 30%.

Each pre-sawed microscanner chip is then separated by careful cleavage. The device is ready for bonding and packaging.

3.2 PROCESS IMPROVEMENTS

3.2.1 Dry mirror release

In order to improve fabrication yield, efforts were concentrated on the microscanner releasing step. The simplest way to achieve a reasonable output was to avoid the wet etch step and to investigate a dry etching method. The release step of the moving part finally consists of a two step process, combining wet and dry etching.

In the first step, KOH attacks the silicon, underetches the structure and liberates the large mirror surface, which lies on the sacrificial layer. This is done without releasing the actuator beam (Fig. 3.3 c) so that the structures remain in the substrate plane. A gap of about 20 µm between the mirror plate and the substrate is created. This gap is large enough to avoid the mirror sticking to the substrate. In the second step, an isotropic plasma dry etching of silicon by SF_6 underetches the actuator beam, releasing completely the mirrors from the substrate (Fig. 3.3 d and 3.4). Besides the fact that this technique is free of surface tensions, the Si / SiO_2 etch ratio (about 2300) is much higher than in KOH (250 in KOH 40% at 60°C), resulting in stiffer bimorph actuators (Cr / SiO_2), higher resonance frequencies but also less sensitive devices.

The combination of these two steps avoids the delicate freeze-drying procedure after wet etching. The dry etching raised the yield considerably up to 95% (Fig. 3.5 and 3.6) compared to the 30% achieved before.

3.2.2 Mirror stiffness and flatness improvement

Mirror curvature caused by actuation or stress gradients inherent to the fabrication process will severely degrade the scanning resolution. Several solutions where tested to improve flatness and to increase plate stiffness. They

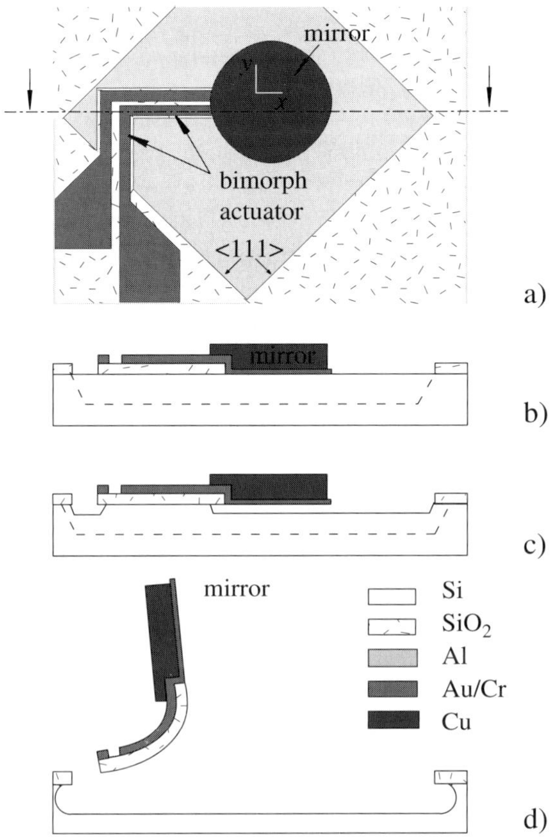

Fig. 3.3 Schematic of the 2D microscanner, type 2 (see chapter 5):
a) top view,
b) cross-section of the device before releasing the moving part,
c) cross-section after releasing the mirror and creating a 20 μm gap between the mirror and the substrate,
d) cross-section after releasing the moving part.

consisted in removing stressed layers, testing thermal treatments and planning reinforced mirrors.

Improving flatness

The first generation mirrors presented irregular and irreproducible curvatures (Fig. 3.7) mainly due to the stress difference between the mirror layer components and the influence of the massively stressed Au / Cr layer (defining the conductor layer on the bimorph as well as the mirror plating base). The latter

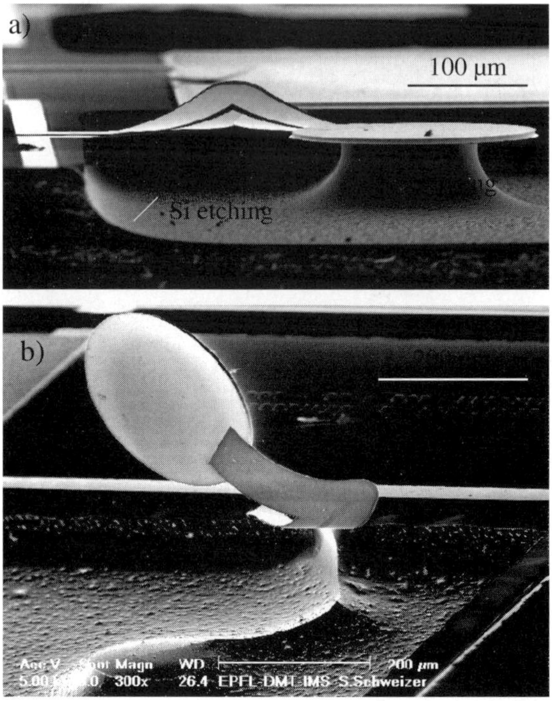

Fig. 3.4 Dry etching of microscanners (SEM pictures). a) Mirror sticking on the silicon if the gap between substrate and mirror is not large enough. Note that the bimorph beam is already completely released. b) Completely released microscanner.

is removed under the plated mirror by adding a mask and a photolithography step (after mask level 2, Fig. 3.1 e), replacing it by a thin 30 nm Au / 30 nm Cr base plating film. This improvement resulted in regular and reproducible curvatures.

Stress in electroplated Cu can easily be relaxed by an adapted thermal treatment [107, 160]. Before releasing the mirror, a careful Cu annealing under forming gas (N_2, H_2) at 300 °C is done, so that Cu does not oxidize. This relaxes the internal stresses introduced during the electroplating of the Cu layer. With this method, the mirror curvature has been reduced by a factor of two. Moreover, measurements showed that the annealing homogenizes the curvature of the mirror, generating a reproducible small curvature.

Radius of curvature determination by diffraction pattern analysis

The method consists of illuminating a mirror with a collimated HeNe laser beam and to measure and analyze the spot reflected. Under a known geometry,

Fig. 3.5 SEM picture of a wafer showing a series of 2D mirrors (type 2). The mirrors have a diameter of 300 μm with a 50 μm by 100 μm long "L"-beam.

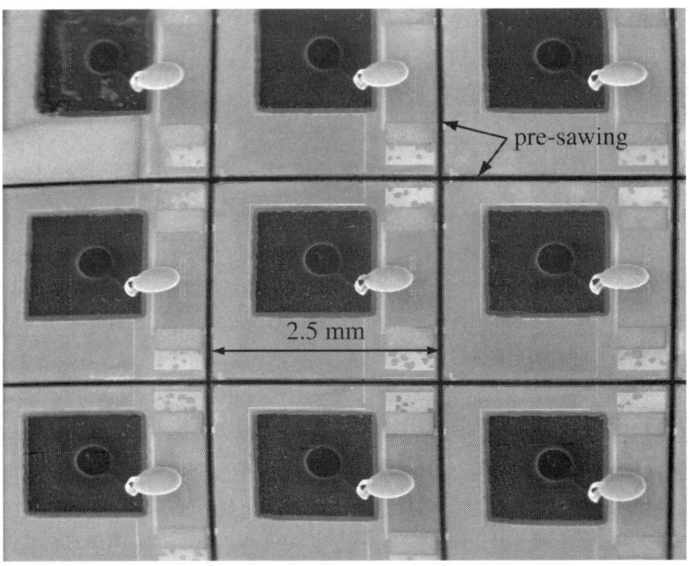

Fig. 3.6 SEM picture of a wafer, top view. The mirrors have a diameter of 300 μm with a 50 μm by 300 μm long "L"-beam.

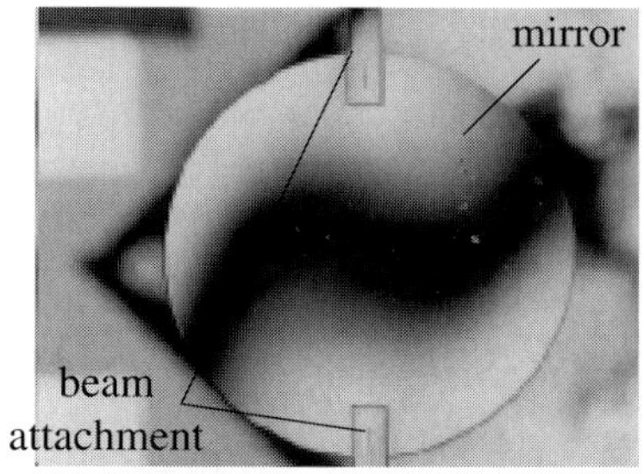

Fig. 3.7 Phase contrast picture (with polarisation) of a first generation mirror (wavelength = 550 nm, Normaski method).

the measurement allows to determine the following data: the rest position angle (Chapter 2) of the mirror, the radius of curvature of the surface for both axes, an eventual tilt of the surface. The method also gives a qualitative indication about the shape and the reflectance of the surface. The measurements were performed by the IRO, EIV (Ecole d'Ingénieurs du Valais), Switzerland [130, 131]. The proposed method should give reliable results for radius of curvature values in the range of 1 to 100 mm.

The measurements were performed for 120 circular mirrors (diameter of 300 μm) on 4 different wafers. Two of the wafers were subjected to a thermal treatment, the others were not. A mean radius of curvature of 11.45 ± 0.55 mm was found for the annealed mirrors in comparison to a mean radius of curvature of 5.05 ± 0.5 mm for the others. For many applications, this is still not sufficient without optical correction.

Improving stiffness

The static stiffness of a micromachined beam or plate is usually restricted to the thickness of the thin-film material. To minimize dynamic deformation of the mirror, reinforced plates have been realized, increasing the bending stiffness of the thin copper layer (Fig. 3.9). Without adding any masks to the fabrication process, the "U"-shaped mirror profile is created by electrodepositing Cu on the whole wafer on top of the structured last mask for mirror electroplating (Fig. 3.8 a). The wafer is then polished, removing the excess material (Fig. 3.8 b). For a 5 μm thick mirror, 8 μm high "walls" improve the stiffness by 20% [161]. It also increases the mass of the mirror by 20%. Experimental measurement showed that the flatness of the mirrors is not improved by this method. A combination of a thermal treatment and a "U"-shaped mirror should be tested in the future. The disadvantage of this method is that the reinforced side

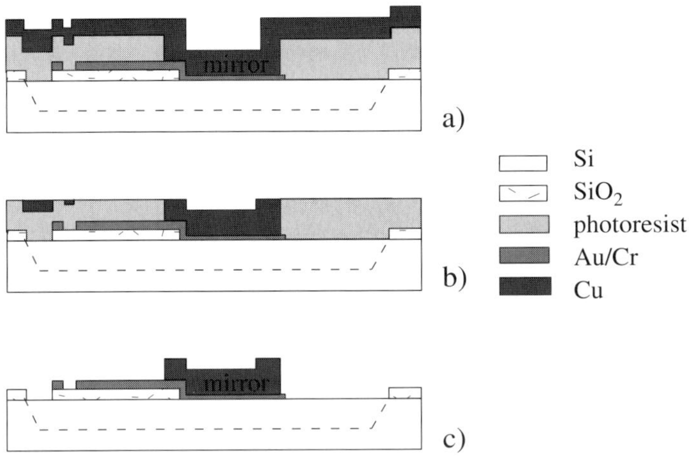

Fig. 3.8 Schematic of the reinforced mirror process. a) electroplated Cu on the whole wafer; b) polishing the wafer, removing the top layer of Cu; c) resist removing.

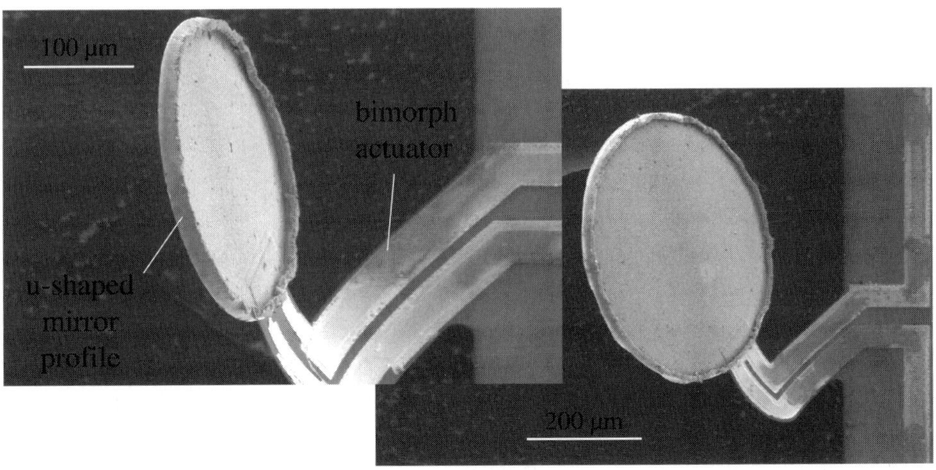

Fig. 3.9 SEM pictures of reinforced copper mirrors.

of mirror can not be used for beam reflection. The “backside” of the mirror has to be used as reflector (Fig. 3.3 d).

An other way to increase stiffness is to use a material with higher Young’s modulus than copper (120 GPa), for example nickel ($E = 210$ GPa) which can also be easily electroplated. Experimental tests were done. However, electroplated Ni layers present more stress than Cu layers and, in addition, are difficult to anneal.

3.2.3 Mirror reflectivity

As seen in section 3.1, depending on the initial rest position, the mirror can be a reflective Cr or Cu layer. The normal incidence reflection coefficient is of about 0.65 for Cr and 0.8 for Cu in the case of a 650 nm wavelength light (which is the laser wavelength for barcode readers) [128]. To increase the reflectivity and the surface quality of the layer, Au (reflectivity of 0.92) could be electroplated instead of Cu. At the same time, oxidation issues are solved and Au can also be annealed for stress releasing.

3.3 CONCLUSIONS

The fabrication process is simple, based on original technological steps allowing the integration of the indispensable functions. A rough estimation of the fabrication expenses results in a chip price of less than 1 US$, for a small serie of about 10 000 mirrors (with ~500 chips on a wafer). Bonding and packaging are not included in the price. The same run can provide one-dimensional as well as two-dimensional micromirrors.

A good fabrication yield of 95% was achieved by a dry microscanner substrate release. The dry plasma etch allows for larger mirror designs, which were limited due to the previous wet liberation etch.

Special efforts were done on mirror flatness, resulting in reproducible and homogeneous curvatures. This allows us to consider fixed optical corrections for resolution improvements.

4 ONE-DIMENSIONAL MICROSCANNER

Nothing has really happened until it has been recorded.

Virginia Woolf

This chapter describes quantitatively the actuation of the optical scanner as well as of the tunable optical filter, the latter will be described more in detail in chapter 6 - 8. The fabrication technology of both devices is different, but the physics behind it is the same.

4.1 TEST SET-UP

Simple manual measurements of the deflection angle can be done with a laser that points on the filter plate and a curved paper screen with scale to read the reflection angle.

For automated testing a test bench has been constructed where a position sensitive detector (PSD) measures the angle of reflection and optionally also the intensity of the reflected laser beam (Fig. 4.1). Different semiconductor lasers with wavelengths of 635 nm (red) and 532 nm (green) have been used. The PSD is a 37 mm long photodiode with two anodes, one at every end. The ratio of the two photocurrents gives the position of the lightbeam (or its center of gravity if it is blurred out), the sum of the photocurrents gives the intensity. The analog design allows an acquisition rate of 100 000 beam positions per second, which would be hardly possible with a CCD detector. A PC with AD/DA card and additional electronics drives the actuator and displays and stores the measured data.

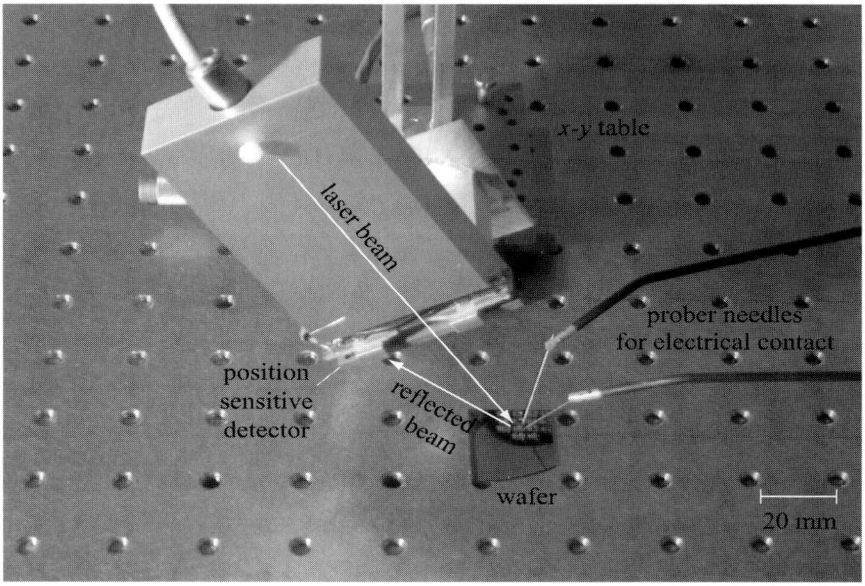

Fig. 4.1 Test bench for measurements of the deflection angle of micromirrors with a laser and a position sensitive detector (PSD).

4.2 Static characterization of Chromium based actuators

4.2.1 Single device characterization

The bimorph, free standing cantilever is heated by an electric current through the conductive metal layer. A given DC input voltage is applied on the actuator and the current and the deflection angle are measured. The behavior of the actuator is linear with the power as the temperature increases with the heating power. By the same way, it is quadratic with the current and with the voltage, at least for small temperature changes that do not affect the heater resistance.

In Fig. 4.2 the optical angle was plotted as a function of the applied power for one device. Sixteen series were done chronologically. Between two series, the device was operated at its resonance frequency during several hours, or it just rested. In some cases, the measurements were done subsequently. As we apply a given voltage, when the average resistance decreases from one series to the next, the sensitivity of the actuator increases. Indeed, the average electrical resistance of the beam decreases and stabilizes (Fig. 4.3 a), explaining the apparent

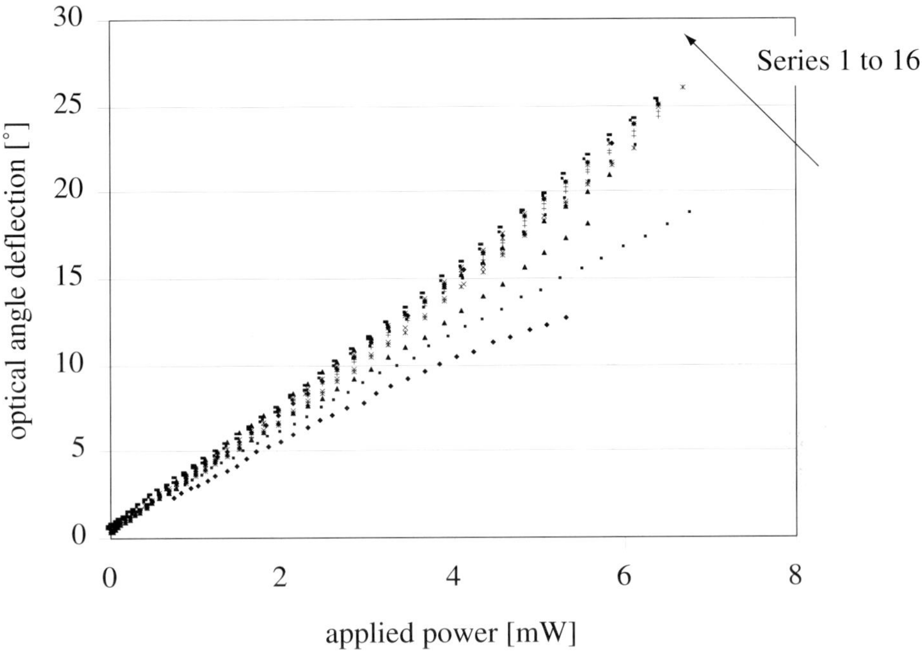

Fig. 4.2 Optical angle vs. applied power (for a given input DC voltage). Between two chronological series, the device was at resonance, made a break, or the measures were done subsequently.

sensitivity increase. When the actuator is heated up, there is an initial "burn-in" phase were the sensitivity of the cantilever varies until it steadies. This is expressed by a decrease and stabilization of the average resistance and by a variation and a stabilization of the initial rest angle (Fig. 4.3). The average resistance decreases from 40 Ω to about 30 Ω (the "burn-in" phase increases the electrical conductivity) and the rest position angle varies about more than 10 degrees (optical angle).

The position detector used has a relative resolution of about 0.1°. For an absolute position measurement, the main source of error is given by the detector dependence on the background illumination. We found an average dispersion of the absolute angle of about maximal ±2° (optical angle).

A linear approximation of the power dependence gives us a maximum average error of about ±0.5 % for the optical angle. One can predict the mechanical deflection angle in a range of ±0.25 % for a given power.

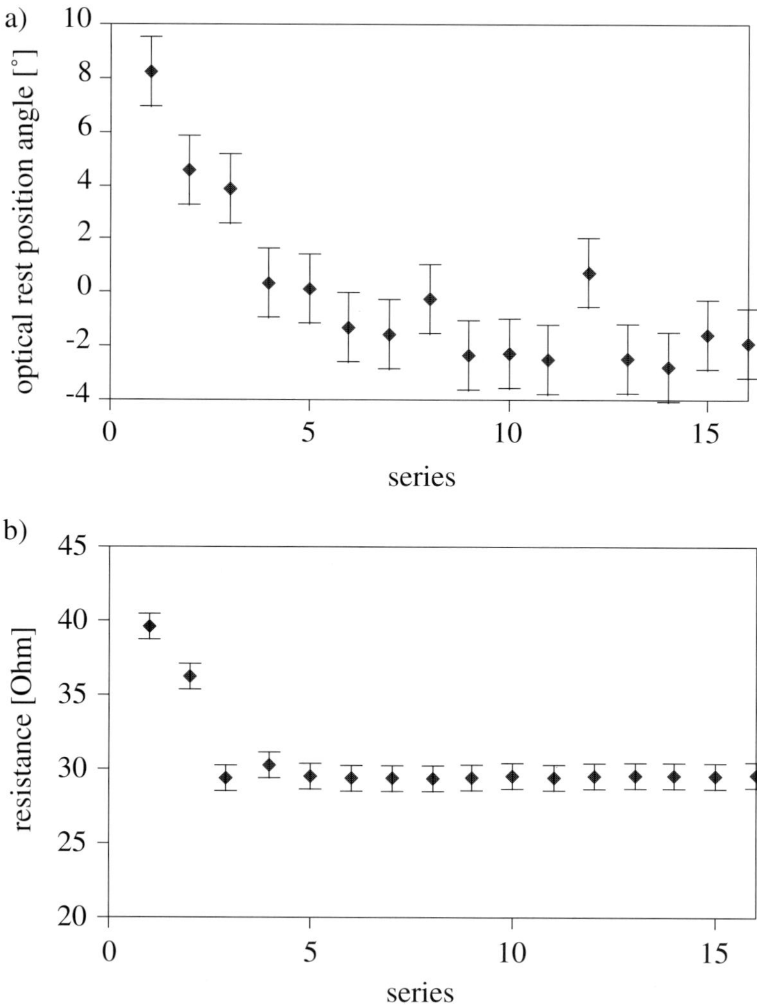

Fig. 4.3 *a) Rest position angle: chronological evolution for the previous series (Fig. 4.2); note that the absolute angle decreases because the beam unfolds after the burn-in phase.*
b) Average resistance evolution for each series.

4.2.2 Comparing angular deflection among various devices

We processed several devices with different beam configurations and different mirror sizes and shapes. The thickness of the beams and length were the same for every device (same wafer). A single beam and a double beam layout were designed (Fig. 4.4). On the single beam layout, the electrical path comes

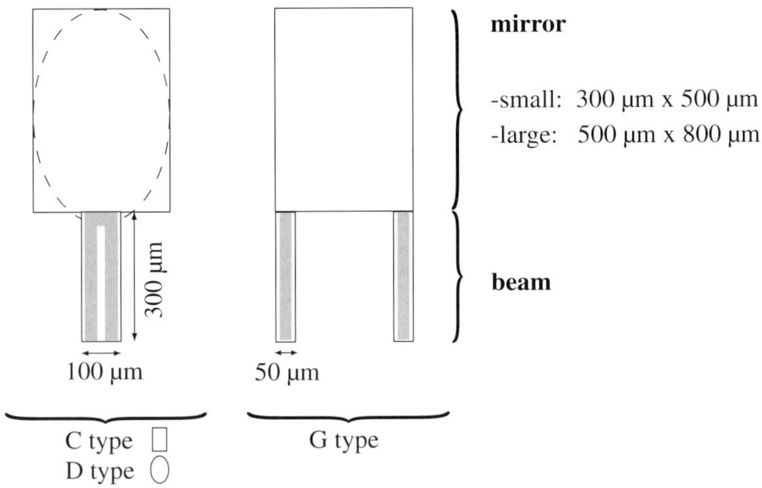

Fig. 4.4 Schematic of the two 1D microscanner configurations.

forth and back on the same beam; whereas on the double beam configuration, the electrical path goes forth on one beam and back on the other. The current passes via the conductive mirror. The measurements presented previously were done for a small mirror with a double beam configuration (G type).

Fig. 4.5 depicts the optical angle as a function of the applied power (a) and the heating current (b) for several small mirror of type G, after "burn-in" procedure. For Fig. 4.5 a, one can observe a variation of the resistance from one device to the other (more than 10%, 3 to 5 Ω for an average of 30 Ω), thus a sensitivity modification in reference to power. A contact resistance of 3 Ω would explain the variation but it is unlikely. The reason of the variation could be related to process conditions which cause a resistivity difference from one point to the other on a wafer [125].

If plotted in dependence of the heating current, the optical angle for a family of micromirrors from the same wafer has a standard deviation of about ±5 % (Fig. 4.5 b), as much as the maximal inaccuracy of the test set-up.

Fig. 4.6 shows the optical angle dependency for different types of small micromirrors. The double beam configuration (type G) is plotted, with the simple beam and rectangular mirror (type C) and the simple beam and oval mirror (type D). The double beam configuration is more sensitive than the single beam configuration. As for the former the current passes through the mirror, the heat resistance is different. This explains the sensitivity difference, which is also

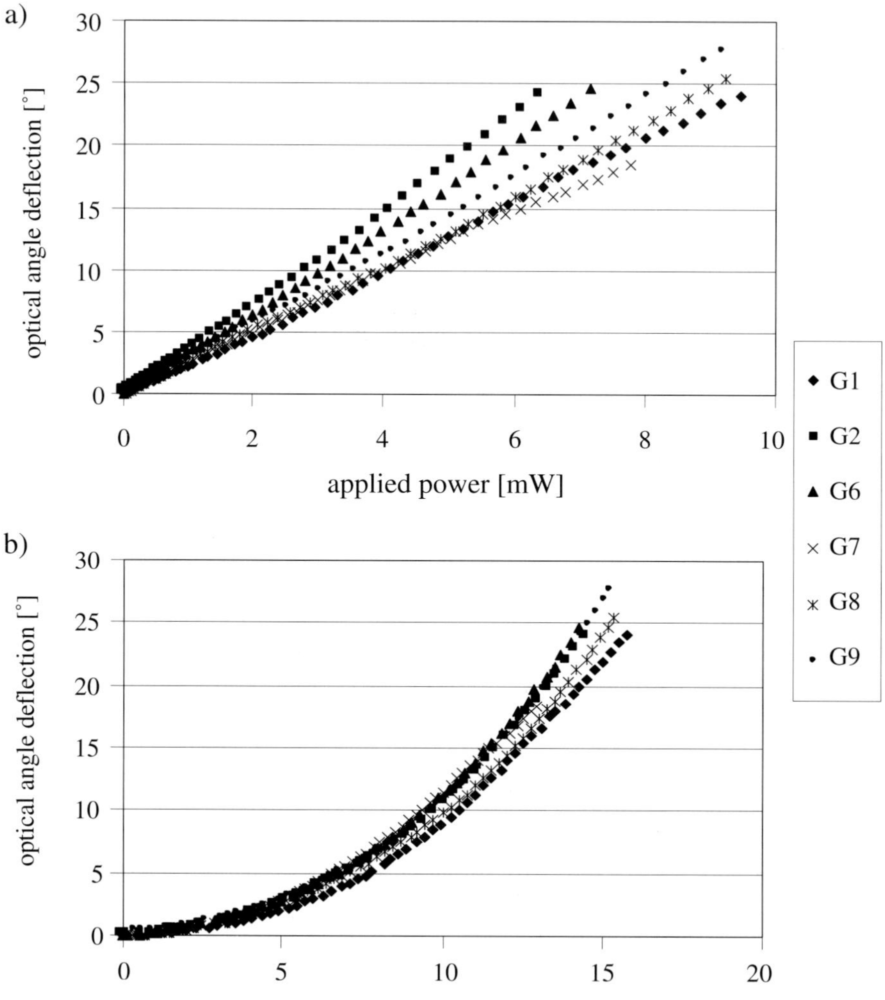

Fig. 4.5 a) Optical angle vs. applied power (for a given input DC voltage) and b) Optical deflection vs. heater current (for a given input DC voltage) for the G-type micromirror from the same wafer. Note that all microscanners were first stabilized by the "burn-in" procedure, which consists of at least 2 hours drive at their resonance frequency.

observed for the large mirrors (Fig. 4.7), but inversed (the double beam is less sensitive than the single beam). For the case of small mirrors, a design inaccuracy in the beam length has been observed such as probably one beam

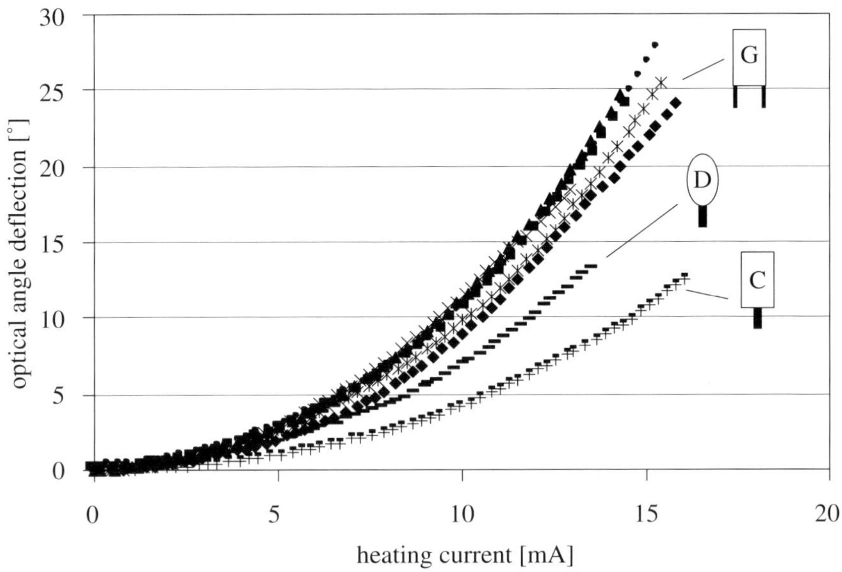

Fig. 4.6 Optical deflection vs. heater current (for a given input DC voltage) for small micromirrors family G, C and D type (from the same wafer); note that all microscanners were first stabilized by the "burn-in" procedure.

heats up faster than the other, or one beam bends more than the other, leading to larger actuator sensitivity.

The oval micromirror (type D) is more sensitive than the rectangular one, independently of mirror size. This phenomenon can be explained by a lower heat loss by the mirror.

There is no major difference between large mirror and small mirror sensitivity, only for the G type (double beam configuration), where the reason is geometrical as the currents path depends on the size of the mirror.

Finally, the presented types of microscanners are capable of maximal optical angles of over 30° (for an average DC power of 8 mW, an average DC current of 16 mA or an average DC voltage of 0.5 V) with an elastic behaviour. Above these values, the curves flatten out due to creep and plastic deformation.

The rest position depends slightly on the environment temperature. This can be compensated elegantly by a drive electronics that works in a constant resistance mode, like it is used sometimes for heating lamps. This mode keeps the heater at a given absolute temperature.

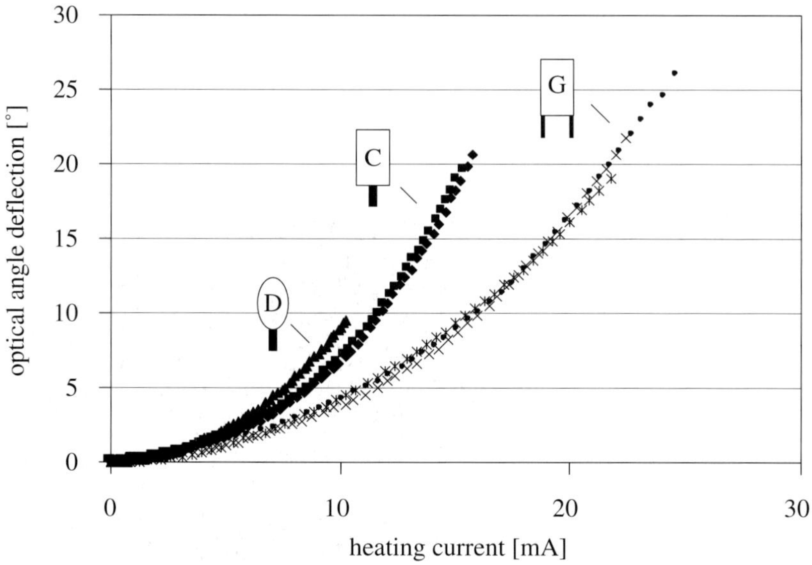

Fig. 4.7 Optical deflection vs. heater current (for a given input DC voltage) for large micromirrors family G, C and D type (from a same wafer); note that all microscanner were first stabilized by the "burn-in" procedure.

4.2.3 Resistance variation of Cr layer with temperature

As already seen in the previous section, the resistance of a device tends to decrease and stabilizes after several drive-in cycles. However, the mean resistivity remains 10 times higher than for bulk material (about 130 μΩ·cm instead of 12.7 μΩ·cm). This is not surprising, as it is well-known that resistivity of thin films is always higher than that of bulk material because of the small grains the thin films are made of, and the numerous resulting grain boundaries that increases the mean resistivity of the layer.

Moreover, as usual for metals, the resistance of the bimorph beam shows a linear variation with the temperature (it increases with temperature). TCR (Temperature Coefficient of Resistance) measurements can give an indication of the perfection of the film material. Often it is much less in thin films compared to bulk material. A small TCR is related to a poor quality thin film with high resistivity, which is our case.

The electrical resistance of metals is generally the sum of resistance resulting from phonon scattering, which is temperature dependent, and resistance from scattering at crystal defects, which does not depend on temperature. The second contribution can normally be neglected for bulk material at ambient temperature, but not for thin films. The resistance temperature dependence of the metal is given by

$$R(T) = R(T_0) \cdot (1 + \beta(T - T_0)) \tag{4.1}$$

where the TCR, here called β, is 0.003 K^{-1} for bulk Cr [128].

For thin films the resistance contribution from crystal defects can be added as a temperature independent constant. It lowers the relative coefficient β compared to bulk material. For our devices, an average β of $4 \cdot 10^{-4} \pm 1 \cdot 10^{-4}$ K^{-1} is found. It is observed that resistance variation with temperature, besides being quite stable and linear within a few percent over the entire range for one device, varies from one device to the other. In other words, the TCR is different when measured and calculated for micromirrors of a same family. The temperatures reached in the beam remains quite low and the normal DC driving interval is between 20 and 200 °C.

Summarizing, we have a variation of resistivity on the evaporated Cr layer giving a variable absolute resistance from one micromirror to the other. This issue can be strongly ameliorated by a careful deposition parameter control especially concerning the pressure and a stable deposition temperature. At the same time, there is TCR variation for a batch of the same microscanner type probably also due to the resistivity variation throughout the layer and the poor microstructure quality of the Cr layer. The TCR combines effects found in the film, i.e. metallic conduction in the grains (columns), with tunnelling through oxides at columnar boundaries.

A possible alternative to the Cr layer for a better microstructure homogeneity, thus electrical stability on a wafer, is the use of Ni or Au with high and continuous layer quality characteristics [125]. In addition, these metals show higher coefficient of thermal expansion (about $13 \cdot 10^{-6}$ K^{-1}). These layers are also known to be very stable electrically showing a good resistivity uniformity.

As the TCR is stable for a "burned-in" calibrated device, a simple resistance measurement can be used to determine the average temperature and the deflection angle of a given microscanner. The advantage is that this makes us independent of the ambient temperature, giving the initial position of the microscanner, for example [261].

4.3 Static characterization of Nickel based actuators

For the above reasons and others (process compatibility), the thermal actuator of the tunable optical filter device (see chapters 6 - 8 for details) was fabricated using a Ni layer instead of Cr. Ni can be evaporated or electroplated, the latter was done here. It needs a seed layer of Cr/Au, which can be neglected in mechanical modelling because of its thickness of only 20 and 25 nm, compared to 800 nm of Ni. The second material for the bimorph actuator is a layer of porous silicon of an effective thickness of 3.0 µm. Further the beam contains a silicon nitride (Si_3N_4) insulation layer in between that can be considered by extending the model to a trimorph actuator, it changes the characteristics only slightly. The cantilevers are typically 100 µm wide and between 270 and 2000 µm long, depending on the design. Most of the results are similar to the Cr actuator, so that only the different properties will be discussed.

4.3.1 Angle vs. temperature

To measure the deflection angle versus a known temperature the device has been heated externally on a hotplate. The results are shown in Fig. 4.8. The deflection angle is almost linear with temperature as expected. An actuation angle of 0.25° per °C was found by curve fitting. For high temperatures the curve flattens slightly. A temperature coefficient of Young's moduli could explain the phenomenon if they are different for the materials used. The higher orders of the thermal expansion coefficient of Ni would increase the actuation angle, so it cannot explain the behavior. It will be seen that this decrease of sensitivity is due to the special geometry of the actuator (see chapter 7).

4.3.2 Resistance vs. temperature

The TCR for bulk Ni is $\beta = 0.0069\ \mathrm{K}^{-1}$ at $T_0 = 20$ °C [128]. The heater resistance was measured on a heating plate with a known temperature and shielding to reduce air convection over it. The TCR was found to be $0.0036\ \mathrm{K}^{-1}$ for an electroplated Ni layer, Fig. 4.9. This can be interpreted in the way that about half of the electrical resistance of the Ni layer results from crystal lattice defects. The specific thin film resistance is about twice the value for bulk Ni, whereas for Cr the thin film resistance was found to be about 10 times higher. The exact values depend on deposition process parameters.

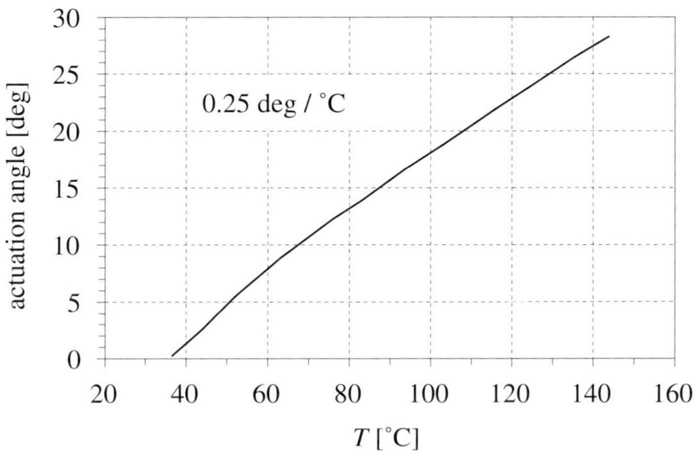

Fig. 4.8 Actuation angle of the thermal trimorph actuator versus temperature on a heating plate.

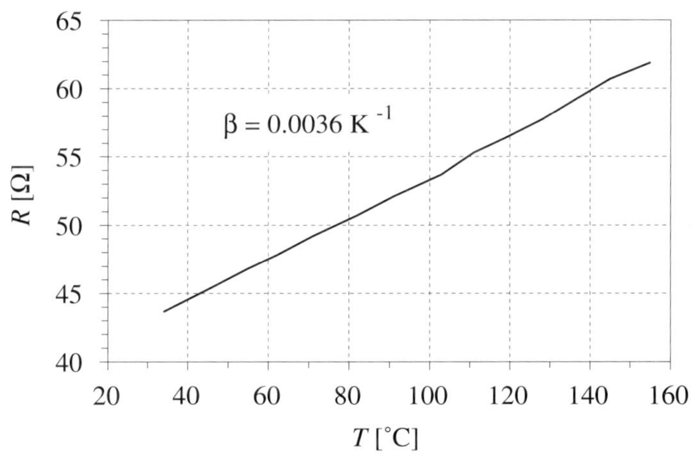

Fig. 4.9 Temperature dependence of the heater resistance, measured on a heating plate at known temperature to determine the temperature coefficient of resistance β. Higher temperatures are more difficult to control precisely because of convection over the heating plate.

4.3.3 Angle vs. power

As the filter plate that is bigger than the micromirror, the actuator has to be is thicker, and the achieved actuation angle per power is only 1.0° / mW. The maximum long term operating power for this device is 80 mW, corresponding to a temperature of about 400°C, considerably more than the Cr based device. If operated at higher power the Ni surface oxidizes and becomes dark. At 200 mW the heater starts to glow red. At this temperature, lifetime is limited to some seconds. The hottest point is found at one quarter of the arm length from the top end. If cooling was purely due to conduction into the substrate the hottest point would be at the tip. The observed temperature distribution allows the conclusion that about three times more heat is conducted into the substrate than into the filter. Convection plays an additional role, especially for long cantilevers, as seen in chapter 2.

4.3.4 Summary

This section discussed the static characterization of the actuator. The actuator reliability depends strongly on the mechanical and the electrical characteristics of the bimorph layer particularly of the conductive layer.

The sensitivity of the actuator (Cr and Ni), its resistance and microscanner rest position stabilize if a "burn-in" is done first, where the device is run at its resonance frequency during several hours.

Besides this, the actuator optical angle precision is given in a range of ±0.5 % for a given current. With the Cr based actuator we achieved typically 20° optical angle deflection for an injected power of 6 mW, thus a current of 13 mA and a voltage of 450 mV. Typical heater resistance was 30 Ω. The devices are capable of maximal optical angles of over 30° with an elastic behaviour.

The Ni based actuator was designed thicker. This makes the cantilever stiffer and decreases the actuation angle per power. However, it can be heated up to higher temperatures, so that the actuation angles are again comparable. Actuators with 1.6 mm long arms and Ni heaters achieved static deflections of up to 90°. The TCR indicates that the electroplated Ni layer contains less crystal defects than evaporated Cr.

4.4 Dynamic characterization

Besides static operation, the cantilever can act as a flexural spring, whereby the mirror constitutes the inertial mass. This permits the device to be used in a dynamic mode where high amplitudes are achieved with little heating power. One of the important elements to be controlled and determined in this system is the mechanical natural resonance of the mirror. The scanner should show high scanning amplitudes, especially for the one-dimensional barcode reader device. These amplitudes must be performed in a frequency range from 50 to 500 Hz.

In order to design the dynamical behaviour of the microscanner and in particular the fundamental resonance frequency of the device, an analytical model and simulations are presented and compared to the experimental results. The analytical model is developed for small deflections. The initial curvature of the device is neglected. It gives the parameters involved, the rules acting on the resonance frequency and has the advantage of being simple.

In order to get closer to the real behaviour, simulations were performed based on a large deflection model, taking into account the initial curvature of the micromirror and the composing layers of the structure. Finally, potential applications are presented, system integration is realized and discussed.

4.4.1 Fundamental resonance frequency calculation

Analytical model

The model developed in this chapter does not take into account the large deflections of the beams that our structure has.

- The clamped beam considered is assumed to be symmetric about the xy-plane (Fig. 4.10). As a consequence, the bending deflections occurs in the same plane, which is known as the plane of bending.
- We consider pure bending, which refers to flexure of a beam under a constant bending moment.
- The initial curvature in the structures is not considered.
- The air damping as well as the thermal behavior of the actuator is not taken into account.

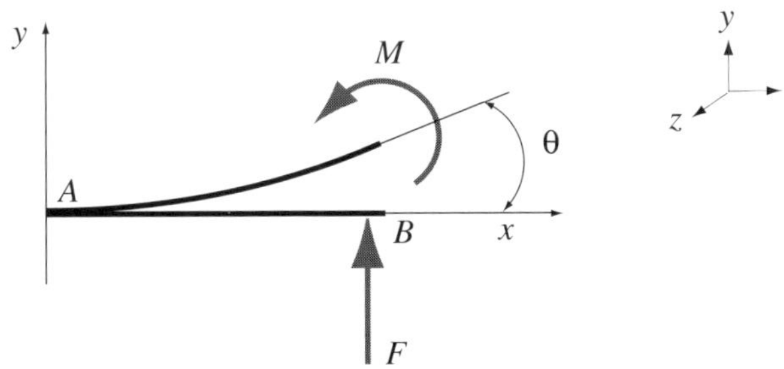

Fig. 4.10 Schematic of the mathematic model of the deflected beam.

A plate (mirror) of mass m_2 and centroidal moment of inertia J about a z axis is attached to the end of a clamped cantilever beam (bimorph actuator) of length l_1, stiffness factor EI, and a mass that is small compared to that of the plate

(50 times smaller than mirror mass in our case, thus negligible). F and M are the resulting force and moment, respectively, acting on the cantilever due to plate mass and inertia. The plate both translates and rotates as, so that it has two degrees of freedom that are defined by the generalized coordinates y and θ shown in Fig. 4.11 [262].

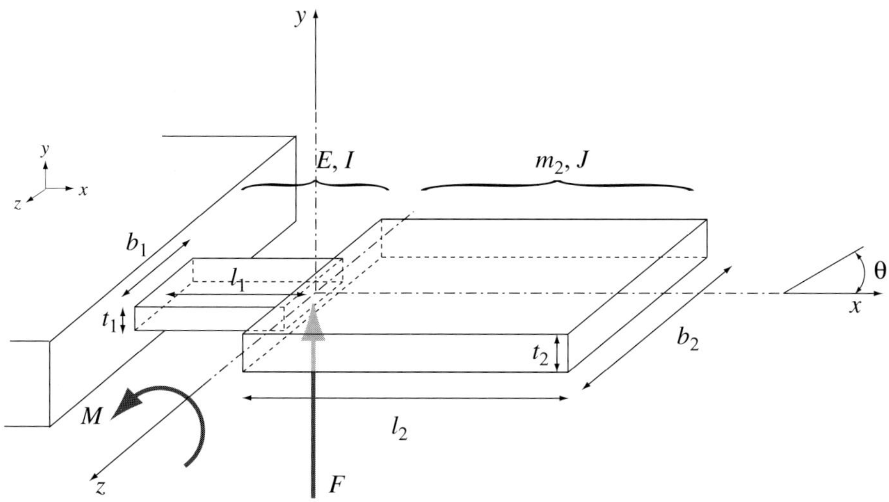

Fig. 4.11 Schematic of the simplified one-dimensional microscanner. The curvature in the beam is neglected.

It is determined from elementary beam theory (considering the assumption mentioned at the beginning of the chapter) that the maximal deflection y_{max} and the rotation θ of the plate due to the application of a force F are [263]:

$$y_{max} = \frac{l_1^{\;3}}{3 \cdot EI} \cdot F$$
$$\theta = \frac{l_1^{\;2}}{2 \cdot EI} \cdot F \tag{4.2}$$

We next apply a positive moment M as shown in Fig. 4.11. The translation and rotation of the plate caused by the application of this moment are:

$$y_{max} = \frac{l_1^2}{2 \cdot EI} \cdot M \qquad \theta = \frac{l_1}{EI} \cdot M \tag{4.3}$$

For a bimorph beam with two different materials A and B, the stiffness becomes [255]

$$EI = \frac{b}{12} \cdot \frac{E_A^2 t_A^4 + E_B^2 t_B^4 + E_A E_B t_A t_B (4t_A^2 + 6t_A t_B + 4t_B^2)}{E_A t_A + E_B t_B} \tag{4.4}$$

With only the mass m_2 of the plate to consider, the differential equations of motion for the undamped free vibration system in terms of flexibility coefficients are ($[B]$ is the flexibility matrix):

$$\begin{bmatrix} y \\ \theta \end{bmatrix} + [B] \begin{bmatrix} F \\ M \end{bmatrix} = \begin{bmatrix} y \\ \theta \end{bmatrix} + \begin{bmatrix} \frac{l_1^3}{3EI} & \frac{l_1^2}{2EI} \\ \frac{l_1^2}{2EI} & \frac{l_1}{EI} \end{bmatrix} \cdot \begin{bmatrix} F \\ M \end{bmatrix} =$$

$$\begin{bmatrix} y \\ \theta \end{bmatrix} + \begin{bmatrix} \frac{l_1^3}{3EI} & \frac{l_1^2}{2EI} \\ \frac{l_1^2}{2EI} & \frac{l_1}{EI} \end{bmatrix} \cdot \begin{bmatrix} m_2 \cdot \left(\ddot{y} + \ddot{\theta} \cdot \frac{l_2}{2}\right) \\ J \cdot \ddot{\theta} + F \cdot \frac{l_2}{2} \end{bmatrix} = \tag{4.5}$$

$$\begin{bmatrix} y \\ \theta \end{bmatrix} + \begin{bmatrix} \frac{l_1^3}{3EI} & \frac{l_1^2}{2EI} \\ \frac{l_1^2}{2EI} & \frac{l_1}{EI} \end{bmatrix} \cdot \begin{bmatrix} m_2 \cdot \left(\ddot{y} + \ddot{\theta} \cdot \frac{l_2}{2}\right) \\ \frac{m_2 \cdot l_2}{2} \cdot \ddot{y} + \left(J + \frac{m_2 \cdot l_2^2}{4}\right) \cdot \ddot{\theta} \end{bmatrix} = 0$$

By determining the stiffness matrix $[K] = [B]^{-1}$, the system of differential equations can be written [264, 265]:

$$\begin{bmatrix} m_2 & \frac{m_2 l_2}{2} \\ \frac{m_2 l_2}{2} & J + \frac{m_2 l_2^2}{4} \end{bmatrix} \cdot \begin{bmatrix} \ddot{y} \\ \ddot{\theta} \end{bmatrix} + \begin{bmatrix} B \end{bmatrix}^{-1} \begin{bmatrix} y \\ \theta \end{bmatrix} =$$
$$\begin{bmatrix} m_2 & \frac{m_2 l_2}{2} \\ \frac{m_2 l_2}{2} & J + \frac{m_2 l_2^2}{4} \end{bmatrix} \cdot \begin{bmatrix} \ddot{y} \\ \ddot{\theta} \end{bmatrix} + \begin{bmatrix} \frac{12EI}{l_1^3} & \frac{-6EI}{l_1^2} \\ \frac{-6EI}{l_1^2} & \frac{4EI}{l_1} \end{bmatrix} \cdot \begin{bmatrix} y \\ \theta \end{bmatrix} = 0 \tag{4.6}$$

with

- $J = m_2 \cdot (l_2)^2/12$ (the contribution of t_2, mirror thickness is negligible) [266].
- $I = n \cdot b_1 \cdot t_1^3/12$ (with n = number of beams), where t_1 and b_1 are beam thickness and width, respectively (Fig. 4.11).
- E is the equivalent Young's modulus for the bimorph (Eq. 4.4).

By inserting the solutions $y = y_0 \cdot e^{i\omega t}$ and $\theta = \theta_0 \cdot e^{i\omega t}$ in Eq. 4.6 and defining η as:

$$\eta = \frac{l_1}{l_2} \tag{4.7}$$

the lower solution, corresponding to the fundamental natural frequency can be written as follows:

$$\omega_0 = \sqrt{\frac{12 \cdot EI}{m_2 \cdot l_1^3} \cdot \left(2 + 3\eta + 2\eta^2 - 2(\eta + 1)\sqrt{(\eta^2 + \eta + 1)}\right)} \tag{4.8}$$

The resonance frequency varies with $t_1^{3/2}$ and $l_1^{-3/2}$. Beam thickness and length are the two dominant factors. The larger η, the lower the natural frequency (Fig. 4.12). The smaller the mirror, thus mirror length (l_2) and inertia, the higher the resonance frequency. The resonance frequency is a function of $t_2^{-1/2}$ where t_2

is the thickness of the mirror directly related to mirror mass. The resonance frequency scales with the square root of elastic modulus *E*.

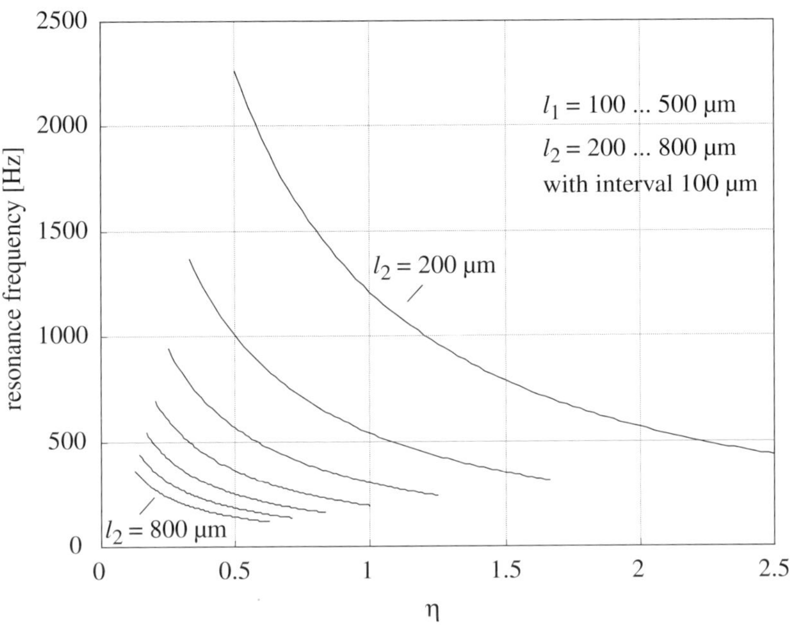

Fig. 4.12 Resonance frequency versus $\eta = l_1 / l_2$. On each line, l_2 is constant and l_1 varies from 100 to 500 μm.

θ or *y* discussed previously (Eqn. 4.2 and 4.3) were obtained by solving the differential equations for a beam with small angles of rotation. When slopes and deflections become large, the exact differential equation of the deflection curve must be used (Appendix A.3). This equation is also based upon the assumption that the material of the beam remains linearly elastic. It is shown that in a DC mode, the small deflection model results in errors (compared to the large deflection value) equal or greater than 10% for mechanical scan angles equal or superior than 18°.

Simulations

As the analytical model does not take into account large deflections performed by the microscanner, neither the curvature of the device, finite element simulations were done in order to estimate the resonant modes of the different structures and to get closer to the experimental results.

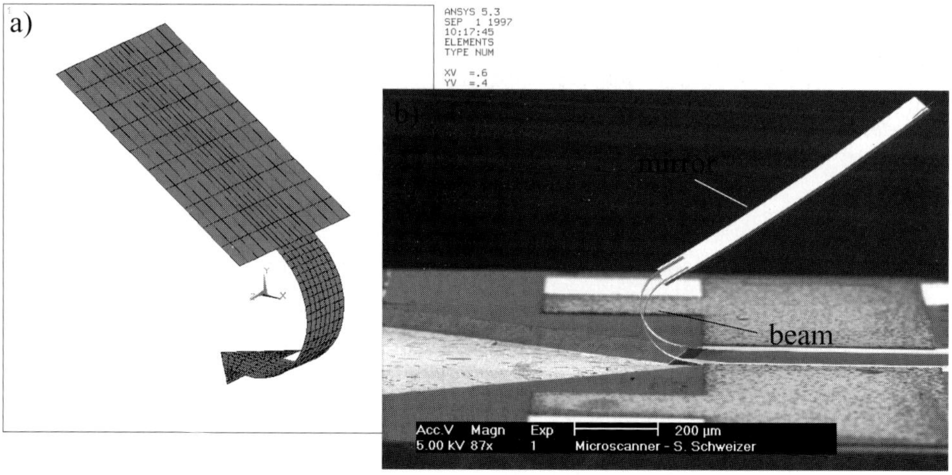

Fig. 4.13 a) ANSYS schematic of the simulated curved microscanner (single beam type). b) SEM picture of a fabricated mirror (500 x 800 μm^2) and a 300 μm long double beam.

The calculations have been carried out with ANSYS 5.5 (Fig. 4.13). The FEM calculations were based on the Euler-Bernoulli beam equations. The device is considered to be a cantilever beam, hence the end of the beam attached to the bulk silicon was completely fixed (clamped beam). The number of elements (between 100 and 500 nodes) were increased until the result reached a convergent value. A simple modal analysis of the scanner, taking into account the bimorph layers and the initial curvature has been done. The initial curvature is imposed in an arbitrary way and is not given by stress causes in the beams. The air damping as well as the thermal behaviour of the actuator has not been taken into account.

4.4.2 Experimental comparison of the resonance frequency

Harmonic AC heating with the appropriate frequency excites a mechanical resonance (fundamental mode), which is situated in the 100 .. 600 Hz range, depending on the geometry of the cantilever and mirror. The movement of the scanning element can be observed with a stroboscope. It appears to be very smooth and regular, independently of the shape (sinus or square) of the applied drive voltage (Fig. 4.14). The single or double beam configurations perform equally well.

A typical scan amplitude versus drive frequency response was measured by reading the deflection of a scanned laser beam on a target screen. We should

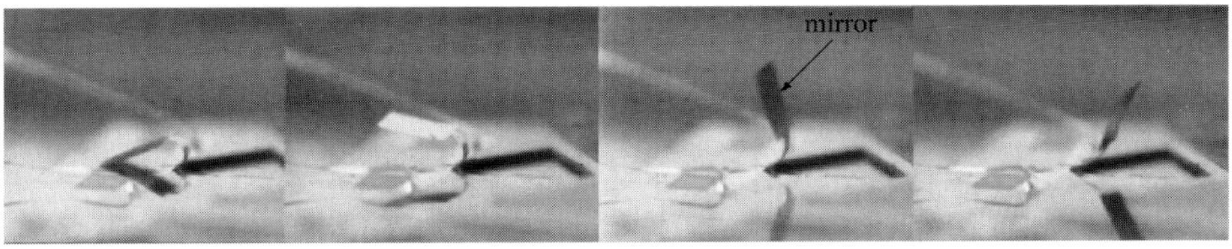

Fig. 4.14 Stroboscopic photos of the device during operation at resonance frequency: 4 mW drive power at 330 Hz (beam length = 300 μm, mirror = 500 x 300 μm², mirror thickness 4 μm).

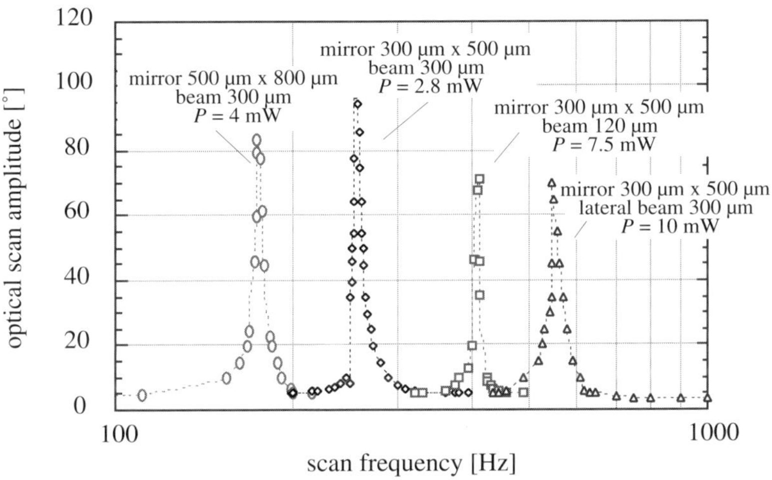

Fig. 4.15 Frequency response (first mode) of different 1D microscanner structures for positive sinusoidal drive (mirror thickness = 6.5 μm). Note that the value on the y-axis represents the optical scan amplitude, which is twice the mechanical deflection of the mirror.

remind that by definition the mechanical scan amplitude, which is the effective deflection angle of the mirror, is half of the optical scan amplitude. The results for different scanner designs are shown in Fig. 4.15. The accuracy of the measurement is ±2°, determined by the angle reading and the laser spot size. A slight resonance frequency shift of about 3.5% has been measured for a same mirror type on one wafer. This is directly related to the thickness of the mirror layer which depends on the resulting thickness precision after mirror polishing (Chapter 3). The polishing results in a thickness precision estimated about 6% between the center and the border of the wafer. As the resonance frequency is a function of $t_2^{-1/2}$ with t_2 being the thickness of the mirror, the resulting imprecision of the resonance frequency can be calculated and equals to 2.8%, which is lower than the resonance frequency shift measured but still corresponding.

Fig. 4.16 depicts calculation (Eqn. 4.8), simulation and measurements (with a standard statistical deviation of ±5% over one wafer) in one graph. The experimentally measured fundamental frequencies are maximally 40% lower than the simulated and the calculated values. The deviations can be explained by differences in assumed and real Young's moduli (these parameters may change by heating), by the neglected air damping effect and the thermal losses. Due to the low layer quality, the elastic modulus of Cr is probably lower than the measured value by nanoindentation (section 2.3.2), resulting in a lower resonance frequency. The assumed clamping conditions may not be precise enough as well. As discussed in Chapter 2, the underetched cantilever results in a new beam length (l_1) definition.

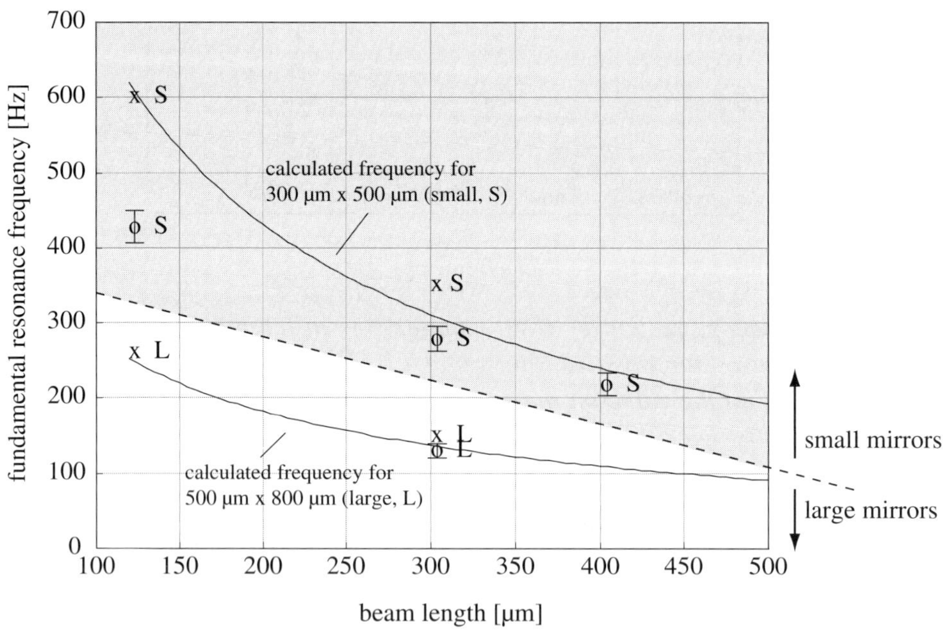

Fig. 4.16 Calculated (lines), simulated (x) and measured (o) fundamental frequency values for two different mirror sizes (thickness of the mirror is 6.5 µm).

4.4.3 Influence of damping

The quality factor Q of the resonating device has a value of about 10 to 20 for the large mirrors and 40 to 50 for small mirrors. It can be determined from the frequency response plot of Fig. 4.16,

$$Q = \frac{f_{res}}{\Delta f_{-3dB}} \Rightarrow Q \cong 10\ldots50 \qquad (4.9)$$

where f_{res} is the resonant frequency, defined as the frequency of maximum amplitude response, and $\Delta f_{-3\mathrm{dB}}$ is the half power bandwidth of the frequency response.

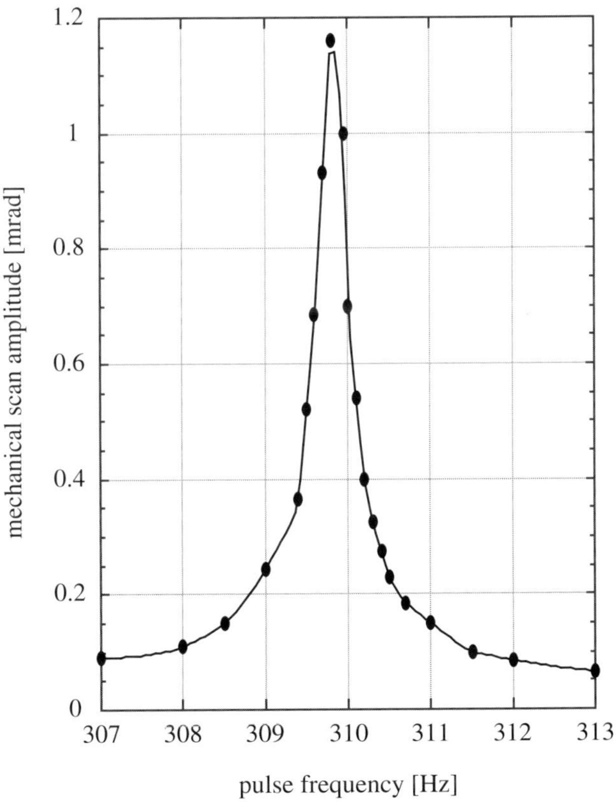

Fig. 4.17 Microscanner resonance curve (dots) recorded in vacuum. The cantilever is excited by a pulsed He molecular beam. The solid line is a fit. In collaboration with A. Wicki et al. [22].

Q is a measure for the energy loss of the resonator. It has also been measured by mechanical excitation of the scanner, where Q is about 70 for small mirrors in air and 1300 in vacuum (see Fig. 4.17) [22]. The difference between air and vacuum indicates that the damping in the surrounding medium (in this case air) is an important factor of energy dissipation.

It has been seen in Chapter 2, section 2.5, that with the "burn-in", the sensitivity of the microscanner increases, the Young's modulus decreases and so does the Q factor. One possible explanation of the lower Q factor (about 50) measured with thermal excitation may be the fact that the micromirrors used for the mechanical excitation (without heat), were not "burnt-in".

4.4.4 Thermal cut-off

The actuator can be heated quickly, but cooling down is the speed limiting factor. The cut-off frequency – at which the amplitude drops by 3 dB compared to DC actuation – varies with the design geometry and material. Values from 9 Hz to 500 Hz were found for long and short beam designs respectively. Fig. 4.18 shows the scan amplitude of the mirror as a function of the excitation frequency. The movement at resonance is symmetrical with regard to the rest position (pointing angle) [267, 114]. The thermal cut-off occurs at a frequency of 50 Hz and the resonance frequency is measured at 330 Hz. This compares reasonably with the theoretical value of

$$f_{cut-off} = \frac{1}{2\pi\tau} \tag{4.10}$$

giving 42 Hz for a τ of 3.8 ms as calculated in section 2.5.2. .

Actuators with heaters made of evaporated Au instead of Cr or Ni showed higher cut-off frequencies at the same geometry. They cool faster because of the higher heat conductivity of the metal tracks.

4.4.5 Dynamic microscanner performances

Staying at the resonant frequency and changing the drive power yields the graph of Fig. 4.19. It shows that the scan amplitude is almost linear with the drive power. At higher powers, however, a slight saturation is observed, which may be explained by the fact that material constants of the metal vary with increasing temperature. The beginning of the saturation occurs for optical angles over 100°, thus inelastic behaviour happens above the required value for our scanning applications. Fig. 4.19 shows that the pointing angle, which is the "average" scan position angle, follows the static sensitivity (about 1.5° / mW).

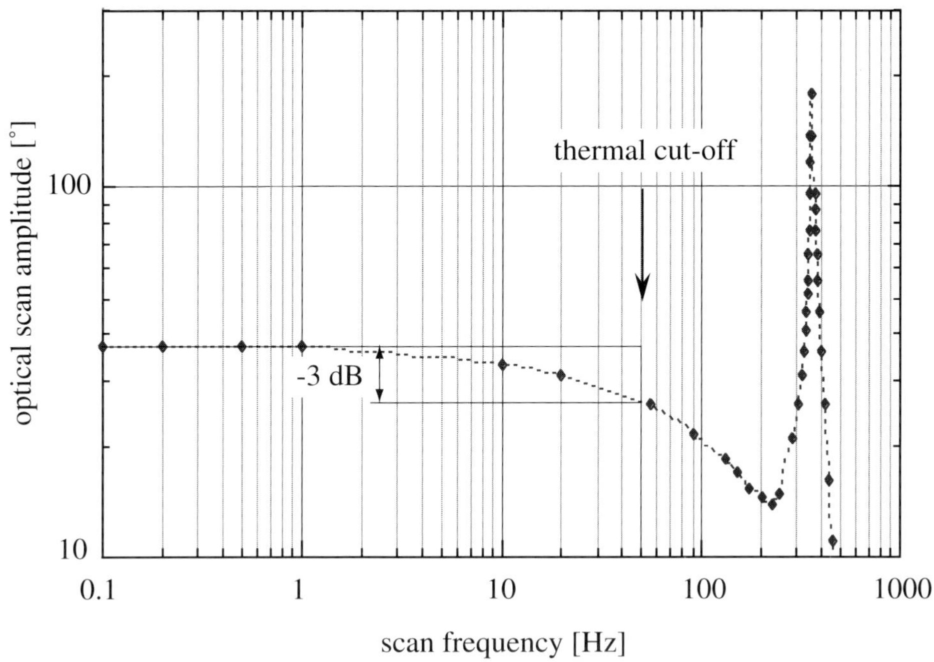

Fig. 4.18 Measured frequency response of a 300 μm long beam with a 300 μm x 500 μm mirror. The cut-off frequency is 50 Hz.

An important issue for scanning applications is the scan repeatability. To determine the accuracy of the microscanner line scan, the position of the scanning laser beam was recorded with a position sensitive detector (PSD). When the scanner is driven harmonically, the scanning laser-beam position can be recorded as a function of time and compared to a sinusoid fit. A standard deviation of about 0.02° over the 15° mechanical scan deflection was measured, thus ~ 0.1% of the full scan.

The oscillator can have some non-linear contributions by the fact that the effective lever length of the cantilever gets shorter when it curls up and longer when it bends down by thermal actuation. The effective spring constant [K] can also vary with displacement following the temperature variation. A simple way to verify if non-linear conditions are present in our system is to plot the frequency response for different average drive power. Fig. 4.20 shows that the resonance frequency tends to decrease slightly (less than 0.3%) with increasing injected power (from 0.06 to 0.8 mW). In this case, no offset power was applied. The same phenomenon can be observed when DC offset power varies.

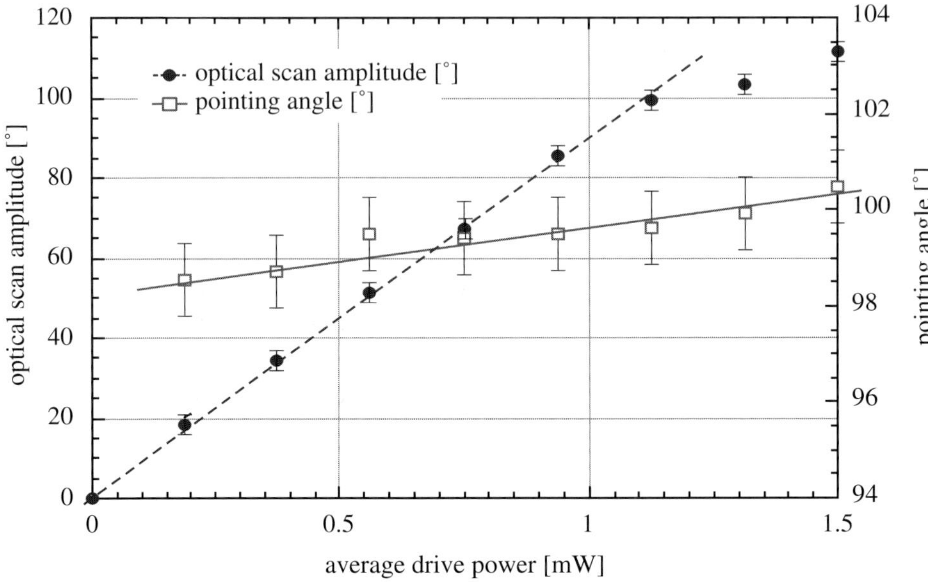

Fig. 4.19 Optical scan amplitude versus drive power at resonance (beam length = 300 μm, mirror = 500 x 300 μm²). Higher angles (up to 200°) could be observed, but were difficult to measure accurately with our setup.

This small deviation can be explained by a temperature increase in the beam, thus a reduced beam Young's modulus and a lowered stiffness. In addition, the frequency response remains symmetric expressing that we have a harmonic oscillator.

In order to demonstrate the robustness during operation at resonance, the device was run for billions of cycles with mechanical deflections of over 90° at resonance (330 Hz). Fatigue tests show high robustness. A resonance frequency shift (of about 3%) is observed, probably due to a mechanical hardening of the metal layer up to 1.5 billion (10^9) cycles at 4 mW (Fig. 4.21), then the device is stable. In the same way, it could be observed that the scanning amplitude stabilizes after a billion cycles as well as the pointing angle.

4.4.6 Lateral suspension mirror

A different mirror configuration has been designed with a laterally attached beam. In order to force rotation about the center of the mirror so as to have a more symmetric movement and to obtain higher resonant frequencies for larger mirrors, laterally fixed beams have been designed (Fig. 4.22).

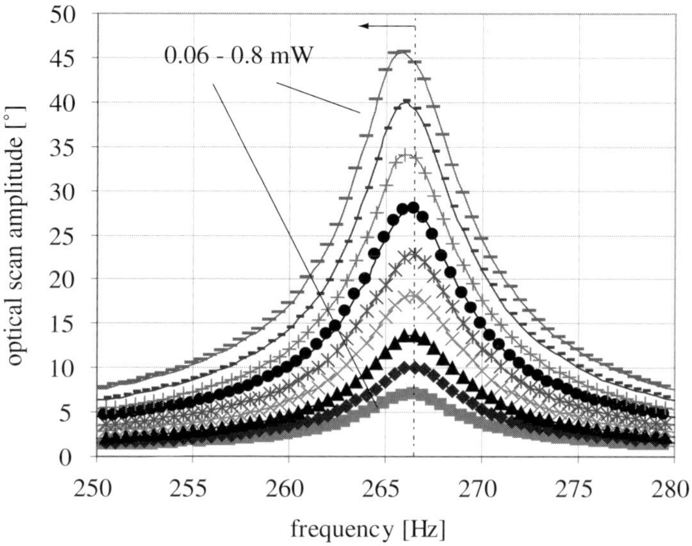

Fig. 4.20 Frequency response of a small mirror microscanner for different average drive power (the system is excited harmonically). No DC power was applied. The position sensitive detector measures the deviation of the reflected laser beam on the micromirror.

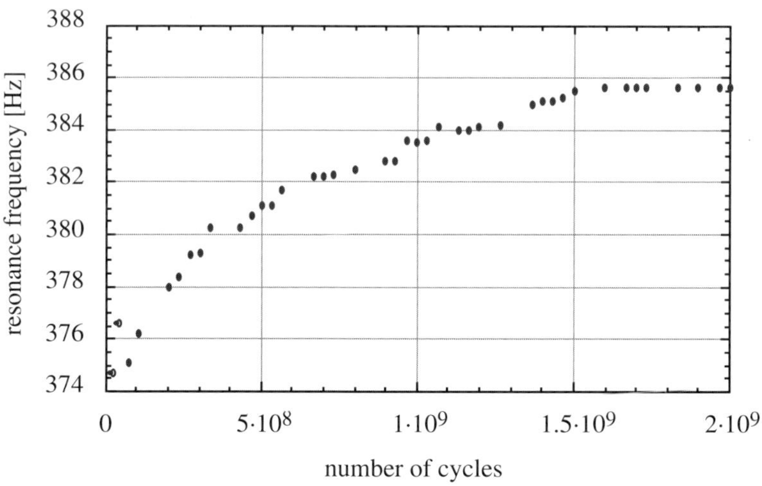

Fig. 4.21 Resonance frequency evolution in the first 60 days of a fatigue test. The mechanical scan angle was 90°. (beam length = 300 μm, mirror area = 500 x 300 μm^2).

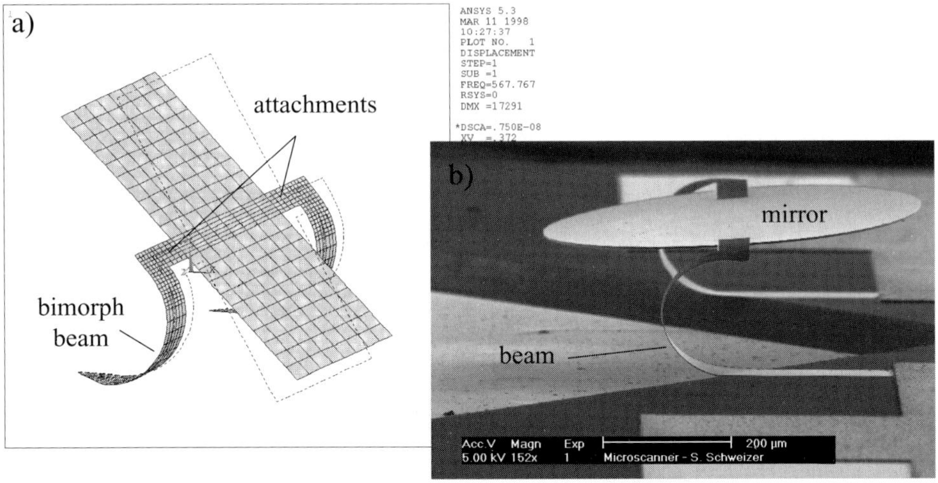

Fig. 4.22 a) ANSYS schematic of the simulated curved lateral beam micro-scanner (rectangular mirror). b) SEM picture of a fabricated mirror (ϕ 600 μm) and a 300 μm long double beam.

An analytical model based on the same considerations as for the cantilever beam scanner was developed. The torsion of the second part of the beams (attachment parallel to the mirror) was not taken into account. The model is developed in Chapter 5 (2D mirror) and is not repeated in this section. Only the results are discussed and presented here (Table 4.1).

One can see that the analytical model has the same type of deviations to simulations and measurements as for the simple beam configuration. Besides the reasons for deviations mentioned in the last section, one can add the fact that the influence of the torsion on the deflection of the small beam is neglected. As for the last mirror type, the beam thickness and the beam length (Fig. 4.22 and 4.23) are the two dominant factors in the resonance frequency determination.

4.4.7 Dynamics of the tunable optical filter with Ni based actuator

The tunable optical filter (see chapter 6 - 8) with the Ni / porous Si based actuator can also be operated in mechanical resonance. Table 4.2 lists the measured resonance frequencies for different designs and compares them with the theoretical values.

The Ni layer of the cantilever is 0.8 µm thick. The Si_3N_4 insulation layer has been neglected. For the first four devices with the suspension at the bottom of the filter plate (type A) the effective average thickness of the porous Si layer has

Table 4.1 Calculated, simulated and measured fundamental resonance frequencies for two lateral designs. Mirror thickness = 6.5 µm. The values are represented in Fig. 4.23.

Mirror size	Calculated res. frequency [Hz]	Simulated res. frequency [Hz]	Measured res. frequency [Hz]
500 x 300 µm^2	910	730	550 ± 27
ø 600 µm (circular)	663	620	450 ± 22
typical deviation from the measured value	60%	30%	-

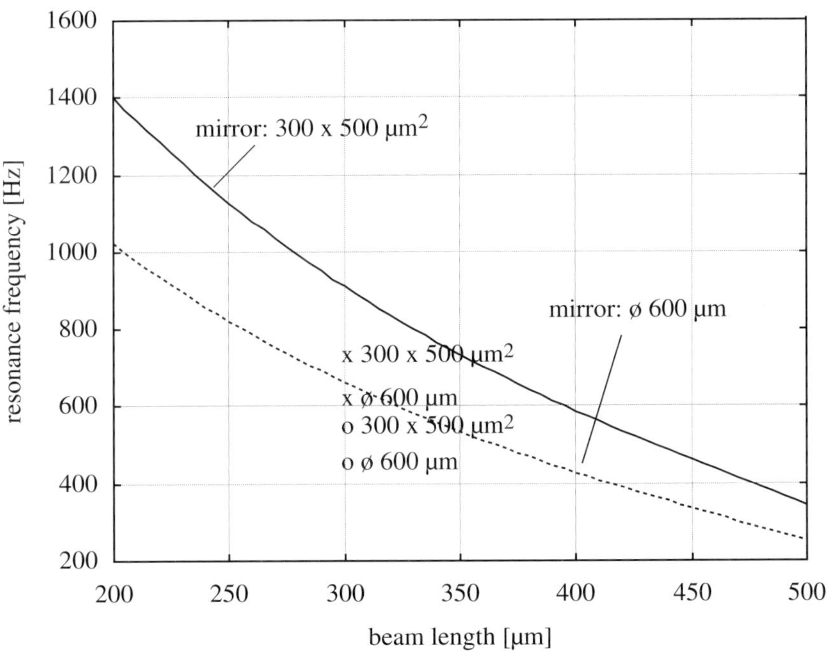

Fig. 4.23 Calculated (lines), simulated (x) *and measured* (o) *fundamental natural frequency values for two different mirror sizes (thickness of the mirror = 6.5 µm).*

been fitted to 4.0 µm. All four devices were fabricated on the same wafer. The analytical theory (Eqn. 4.8) describes well the resonance frequency in

Table 4.2 Fundamental resonance frequencies of suspended filter plates of porous Si for some different geometries and suspension types. The theoretical values are calculated analytically.

suspension type, plate dimensions	cantilever length l	cantilever width b	measured f_{res}	calculated f_{res}	fitted $t_{por\ Si}$
A 410 x 650 μm^2	270 μm	150 μm	1192 Hz	1244 Hz	4.0 μm
A 620 x 820 μm^2	330 μm	240 μm	860 Hz	810 Hz	4.0 μm
A 1580 x 2000 μm^2	500 μm	600 μm	147 Hz	139 Hz	4.0 μm
A 2570 x 3000 μm^2	930 μm	1000 μm	68 Hz	63 Hz	4.0 μm
B 580 x 800 μm^2	720 μm	80 μm	1100 Hz	= 1100 Hz	6.05 μm
B 1100 x 1850 μm^2	1680 μm	80 μm	62 Hz	= 62 Hz	2.95 μm

dependence of the actuator dimensions. The deviations are in the order of the reproducibility of fabrication. The two last filter plates with lateral suspension (type B) have been made on different wafers and thicknesses of porous Si. Here, the measured resonance frequency was used to calculate the corresponding effective thickness of porous Si. With etch parameters to achieve especially thin cantilever arms, a f_{res} of 24 Hz could be achieved with type B.

For the final infrared spectroscopy application the mechanical resonance is of less practical importance. The device will be operated mainly at frequencies of 10 Hz or slower because of the detector response time.

4.4.8 Summary

The combination of analytical calculation and simulation gives a good estimation and prediction of the fundamental resonance frequency, allowing us a reliable design. In particular, a qualitative estimation of the impact of any change in the design on the resonance frequency can be easily done. The beam geometries, especially thickness and length are the main factors inducing changes on the natural microscanner resonance.

Different types of microscanners were fabricated and tested, they are summarized in Fig. 4.24. With the developed microscanners, we achieved large scan angles at a low power consumption. Many scanning applications require optical scan amplitudes of typically 60°, which can be achieved by an average

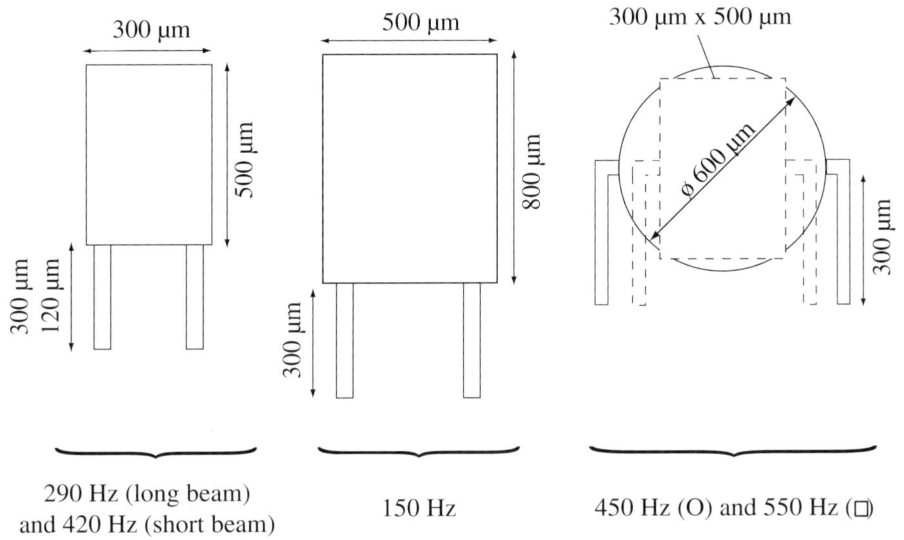

Fig. 4.24 Schematic of the fabricated microscanners and corresponding resonance frequencies for a 6.5 µm thick mirror.

drive power as low as 1 mW. The scanner has a linear harmonic behavior, an excellent scan repeatability and shows high robustness.

The tunable optical filter showed that plates as large as 2570 x 3000 μm^2 can be actuated with a thermal bimorph made of Ni and porous Si. To remain mechanically robust, the beam has to be thicker, which leads at the same time to longer beams, see Table 4.2. With a laterally suspended filter plate of 1000 x 1850 μm^2 mechanical deflection angles of up to 90° have been achieved statically and in resonance. The power consumption is considerably higher, up to 80 mW. Details will be discussed in Chapter 7 and 8.

4.5 1D SCANNER APPLICATIONS

4.5.1 Barcode reader

A possible application of the one-dimensional microscanner is the development of a new type of integrated optical microscanner head working in large scan amplitude at medium frequency. Preliminary studies have shown that the actuator meets the specifications of the barcode scanning application.

Optical design

The target application of the device places some constraints to the configuration of the optical system. Semiconductor lasers or LEDs can fulfil the requirements of miniaturization, low-cost and appropriate wavelength (low optical output power and wavelength in the visible region of the optical spectrum). Numerical simulation using ray tracing allowed determination of the constitutive parameters of the optical system as well as the evaluation of its aberrations. They were performed with the program Zemax. This design depends entirely on the characteristics of the micromirror, which acts as an aperture in the optical system. The resolution of the optics is mainly limited by the size of the aperture, but the aberration of the wavefront and the planarity of the mirror can not be neglected.

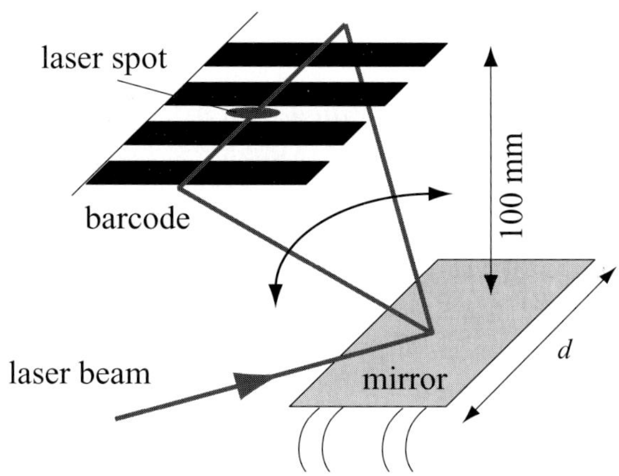

Fig. 4.25 Scanner head constraints.

The smallest distance of two bars in a barcode is about 200 μm (Fig. 4.25). To read the barcodes accurately, an optical laser beam with a spot diameter of 150 μm at 10 cm distance from the scanner head is required. To read also non-planar barcodes, the depth-of-field of the optical system must be sufficiently large (2 cm) such that the spot size at the barcode is always small enough to resolve the thinnest bar.

Simple diffraction calculations show that the current mirror sizes ($d = 500$ μm) satisfy this constraint and that the needed scan amplitude is fully achieved with the actual micromachined scanner design [268].

If the wavefront has aberrations, such as astigmatism due to the laser source and if the mirror is not plane, the laser spot will be larger than the ideal one and thus lead to contrast losses, which could render the barcode scanning unreadable for the electronics. A calculation was developed by the Ecole d'Ingénieurs du Valais, Switzerland, [268] to estimate the light reflected from the barcode taking into account the shape of the mirror and the laser source. It resulted in an impulse signal (light reflected from the barcode) from which the modulation transfer function (MTF) can be extracted (with the program Mathematica) and which is of great interest for the design of the electronic processing.

Vibration sensitivity

The vibration sensitivity s is defined as the displacement divided by the acceleration the device is submitted to. A simple mass-spring model shows that for a 1D device which has a vibration sensitivity s, the vertical deflection y of the mirror can be given as a function of the first resonance frequency and the acceleration for a constant beam length (Eqn. 4.12).

$$f = \frac{1}{2\pi}\sqrt{\frac{k}{m}} \tag{4.11}$$

$$s = \frac{y}{a} = \frac{m}{k} = \frac{1}{(2\pi f)^2} \tag{4.12}$$

where

- f = resonance frequency
- m = inertial mass
- a = acceleration
- k = spring stiffness
- y = displacement

This gives us for typical resonance f = 300 Hz devices, a sensitivity of 0.3 μm/g and for resonance f = 100 Hz devices, a sensitivity of 2.5 μm/g. One can see that for the micromirror configuration we have, small resonance frequencies (about 300 Hz) are not suitable, resulting in large vibration deflection (y is proportional to f^{-2}). This tells us that for low frequency devices, a torsion suspension should be preferred.

One-dimensional scanning system

- *System implementation*

The microscanner (with a mirror size of 500 μm by 300 μm) was implemented into a compact head in which a GRIN (index gradient) lens

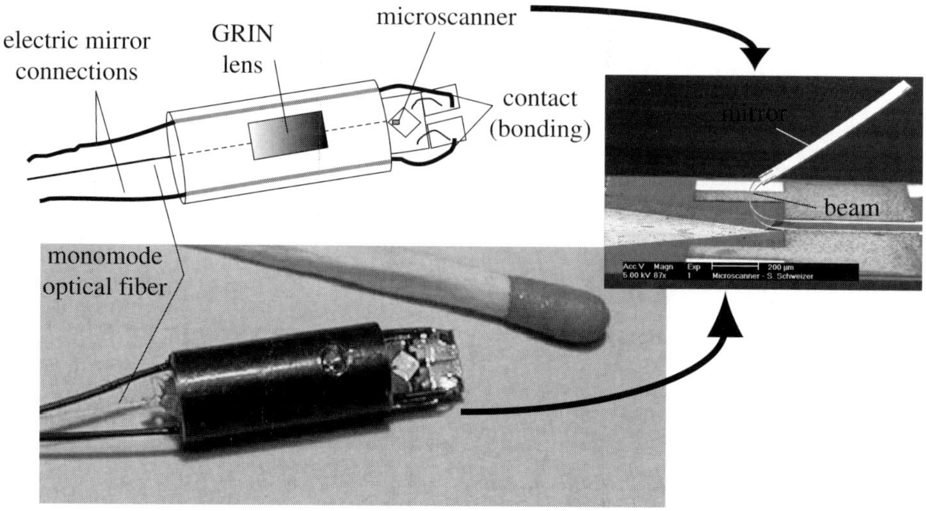

Fig. 4.26 Compact demonstrator for the one-dimensional scanner.

between the output of a monomode fiber and the mirror (with a radius of curvature of 5.8 mm) was adjusted to give a spot as small as possible in the direction of the scanning (Fig. 4.26) and at a distance of 100 mm from the micromirror. A photodiode was temporarily adjusted to this scanner head and a barcode with 1 mm thick rods could be read accurately.

With improved mirrors (annealed) and a sophisticated optical design, we can expect a better spot quality and the possibility to read standard barcodes.

- *Other approaches*

Various scanner head configurations and packaging were imagined. In Figure 4.27, the so-called SOB (Silicon Optical Bench) technology is planed [269], commonly used for optical system alignement on silicon substrates. With this technology, a transversal alignement precision of ±1 µm is achieved. The light that is emitted from a laser diode is collimated by a spherical micro-lens.

Other compact hybrid techniques were imagined where the future use of vertical laser diodes (first prototypes already available commercially at 635 nm) could be of great interest, as presented in the Fig. 4.28. In this case, the PCB (Printed Circuit Board) and the silicon is directly used for optical component assembly.

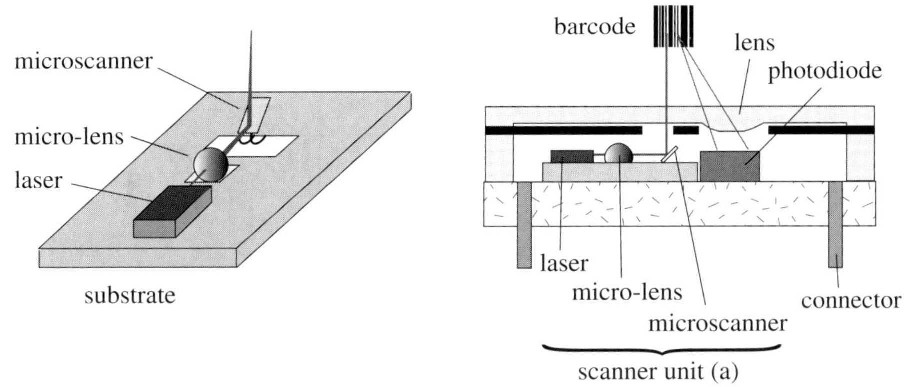

a) scanning unit head b) complete system

Fig. 4.27 Micro-optical bench system based on SOB technology.

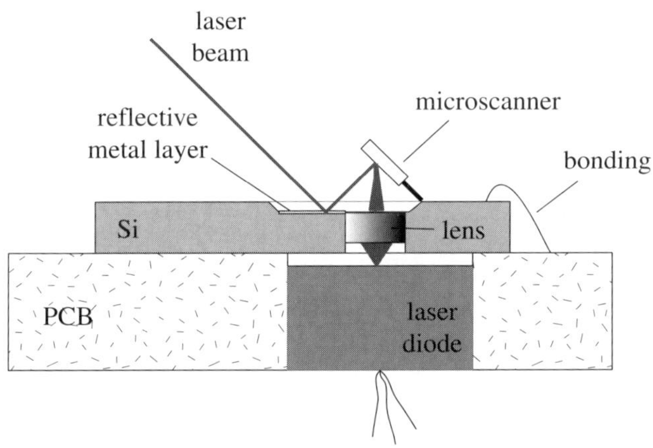

Fig. 4.28 Schematic of a scanning system integration. The lens accommodation could be created by an anistropic plasma etch in the substrate.

4.5.2 Micromechanical detector for molecular beams

The efficient detection of neutral chemically inert beams remains a challenge in experimental physics. Noble gas beams, especially He beams, have important applications, for example, in surface science where they play a role as probe particles comparable to the role of neutrons in solid state physics. They allow, e.g., surface phonon spectroscopy and the non-destructive study of weakly absorbed molecules on surfaces. The most versatile detector with satisfactory

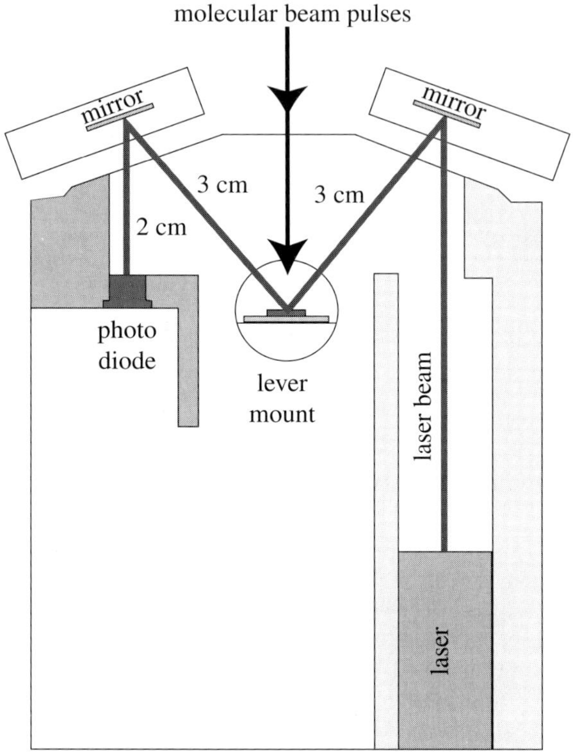

Fig. 4.29 Schematic of the detection setup: base plate and beam deflection geometry for vibrational amplitude read-out.

performance for thermal molecular beams is the electron-bombardment ionization detector, commercially available. Its counting efficiency is, however, low.

The detection of a molecular beam by means of a micromechanical momentum transfer detector operated in vibrational resonance was demonstrated [22]. With a sensitive surface area of 300 x 500 μm^2, the one dimensional microscanner allows us to detect a beam with $6.5 \cdot 10^7$ He atoms hitting the surface per second. The rotational amplitude of the detector is measured by the laser beam deflection method in which the angle is detected by the displacement of a reflected laser beam on a four-quadrant photodiode (Fig. 4.17 and 4.29). The detector response time equals the damping time of the mirror oscillation of about 1 sec. The detector is sensitive enough to measure intensities in molecular beam scattering experiments. Fabrication from single crystalline Si with a smaller sensitive surface and a higher vibrational frequency is expected to improve performance, making the detector more efficient.

4.6 CONCLUSIONS

Various one-dimensional microscanner types have been realized and characterized in a dynamic mode. The devices can be excited at their fundamental resonance frequency between 100 and 600 Hz, depending on their geometries. The parameters involved for the natural resonance frequency are given by a simple analytical model developed in this chapter. In order to increase the accuracy of the design, FEM simulations were carried out, taking into account the initial curvature and the large deflections of the microscanner. The measurements were in good agreement with the simulations.

The scanners, besides showing high angular deflection and low consumption, present a very smooth and regular movement, with good scan repeatability. The devices are capable of mechanical scan angles greater than 90° for frequencies varying from 100 to 600 Hz. Up to now, no other microscanner yields such performance regarding deflection angle. Even other microscanners actuated by a thermal bimorph cantilever [103, 4, 55] maximally reach 20° mechanical scanning angle. The lifetime of the devices goes beyond our expectations. So far, no device ever broke due to fatigue.

The excellent dynamic behaviour of the one-dimensional microscanners opens the possibility to a new generation of scanning heads for a barcode readers. A compact scanning head was implemented and demonstrated. The scanning resolution was not yet sufficient for a standard barcode reading. However, with improved mirror flatness (Chapter 3) and an adapted optical design, a better spot quality can certainly be achieved.

The same actuation principle has been employed in another application to turn an optical interference filter in order to tune its wavelength. This device, which will be described in detail in chapter 7, is fabricated with a slightly different technology, using Ni as heater material. This has some advantages over Cr, like higher thermal expansion coefficient, less lattice defects, and it supports higher temperatures. The analytical model predicted well the necessary dimensions for the actuator to turn the plate, which is much larger that the barcode scanner mirror, at the expense of higher power consumption.

5 TWO-DIMENSIONAL MICROSCANNER

5.1 PRINCIPLE

The two-dimensional microscanner contains an asymmetry in the beam to induce a perpendicular motion. The bimorph beam has an "L"-shape (Fig. 5.1 a). The electrical path is incorporated on one beam. The high sensitivity of the thermal bimorph actuator allows a slow (non-resonant) movement of the scanner in one direction. This movement is mainly created by the large deflection of the long part of the beam and the scanner rotates around the y-axis (Fig. 5.1 b). This movement will be called frame scanning at low frequency. The second orthogonal scanning direction is a resonance motion due to the movement of the short beam. In this mode, the scanner twists around the x-axis and this movement provides raster scanning at high frequency [162, 116].

We realized several types of micro-deflectors [115] for a raster-scan system. Type 1 was the first device processed and has a 500 by 300 μm^2 rectangular mirror, with a 300 μm by 120 μm long "L"-beam (Fig. 5.2 a). Type 2 is the improved version consisting of a 300 μm diameter round mirror with a 50 μm by 100 μm long "L"- beam (Fig. 5.2 b).

5.2 DESIGN AND MODELLING OF THE RASTER NATURAL FREQUENCY

5.2.1 Analytical model

The microscanner will be used in a non-resonant mode for frame scanning and in a resonant mode for raster scanning. For the design, the resonance frequency of the raster scanning at high speed is of greatest interest. From mode shape observation on the constructed microstructures it has been observed that the deflection of the small beam activates a twisting of the mirror around the z-axis (Fig. 5.3).

In a similar manner and for the same reasons as for the one-dimensional scanner, this chapter presents an analytical model and simulation calculations for the determination of the raster scanning frequency. The analytical model based

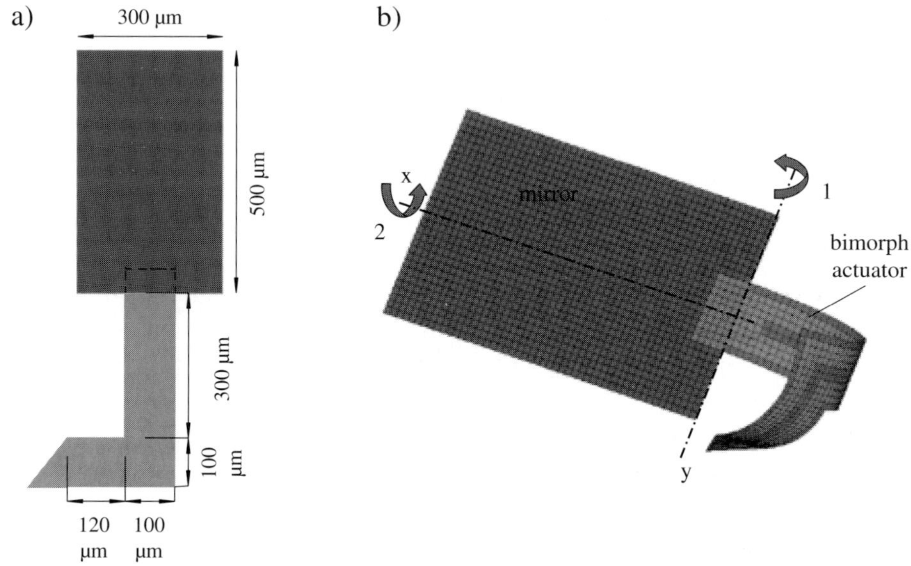

Fig. 5.1 *a) Schematic of the simulated type 1 2D-microscanner (mirror size of 500 by 300 µm² rectangular mirror, with a 300 µm by 120 µm long "L"-beam); b) ANSYS schematic of the simulated released microscanner: the beam curls up, due to the residual stress in the bimorph layers. The two motions are indicated by the arrows:*
1. y-deflection / high sensitivity / frame scanning,
2. x-torsion / high frequency / raster scanning.

on the small deflection theory and neglecting the initial curvature of the microscanner reveals the dependencies determining the resonance frequency. The FEM simulations allow a more accurate modelling including large deflections and initial curvature of the beam.

A simple model was developed for the high resonant frequency determination of the raster scanning, considering the deflection of the small beam (length l_1), with an inertial mirror plate at the end of the long beam (Fig. 5.3). The plate and the long beam (l_1') are considered as an element, which has a general plane motion (it both translates and rotates) as shown in Fig. 5.3, so that it has two degrees of freedom that are defined by the generalized coordinates y and θ. The torsion of the long beam (l_1') is neglected. Based on the same

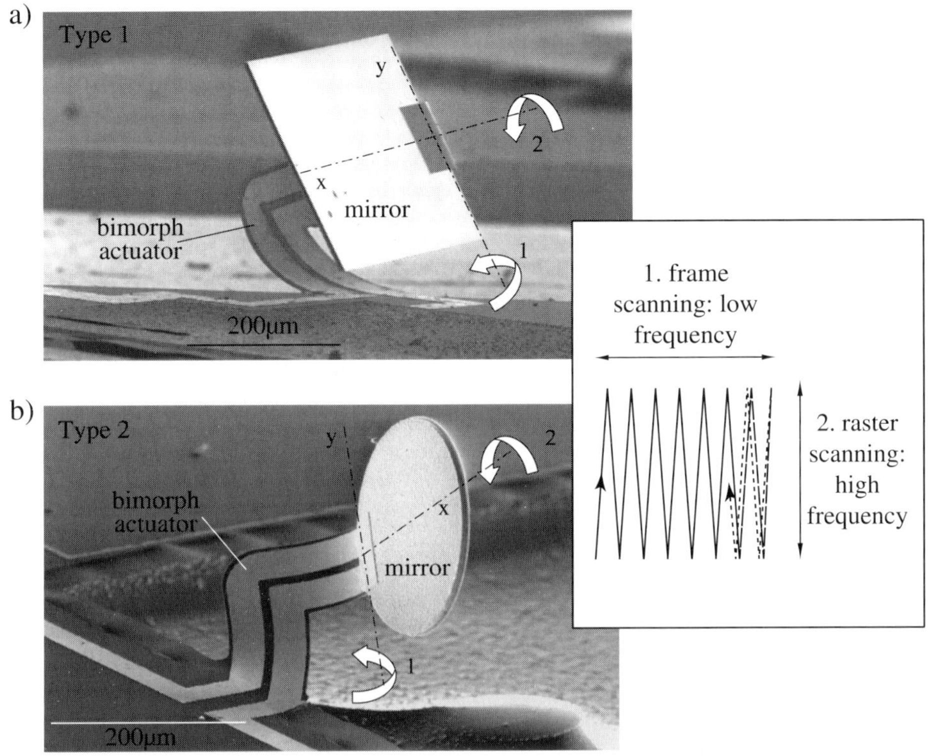

Fig. 5.2 *SEM picture of the released 2D microscanner by selective bulk silicon etching. The two different motions are indicated by the arrows:*
1. y deflection / high sensitivity / frame scanning,
2. x torsion / high frequency / raster scanning.
a) Type 1: mirror size of 500 by 300 μm^2 *rectangular mirror, with a 300* μm *by 120* μm *long "L"-beam. b) Type 2: mirror size of 300* μm *diameter, with a 50* μm *by 100* μm *long "L"-beam.*

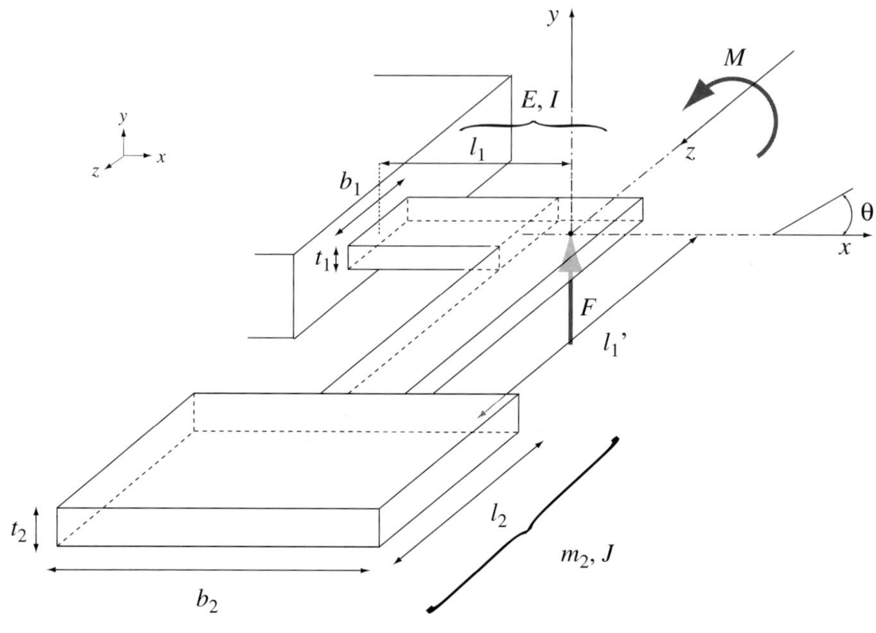

Fig. 5.3 Schematic of the simplified two-dimensional microscanner. The curvature in the clamped beam is neglected.

assumptions as for the one-dimensional micromirror, a system of differential equations can be written:

$$\begin{bmatrix} m_2 & 0 \\ 0 & J \end{bmatrix} \cdot \begin{bmatrix} \ddot{y} \\ \ddot{\theta} \end{bmatrix} + \begin{bmatrix} \frac{12EI}{l_1^3} & \frac{-6EI}{l_1^2} \\ \frac{-6EI}{l_1^2} & \frac{4EI}{l_1} \end{bmatrix} \cdot \begin{bmatrix} y \\ \theta \end{bmatrix} = 0 \qquad (5.1)$$

with

- $J = m_2/12 \cdot (b_2^2 + t_2^2)$ (for the rectangular mirror) [266].
- $J = m_2/12 \cdot (3R^2 + t_2^2)$ (for the circular mirror, R is the radius).
- $I = b_1 \cdot t_1^3/12$, where t_1 and b_1 are beam thickness and width, respectively (Fig. 5.3).
- E is the equivalent Young's modulus for the bimorph (Eq. 4.4, [255]).

By defining η_r and η_r' for the rectangular mirror and η_c and η_c' for the circular mirror as follows:

$$\eta_r = \left(\frac{l_1}{b_2}\right)^2 \tag{5.2}$$

$$\eta_r' = \left(\frac{t_2}{b_2}\right)^2 \tag{5.3}$$

$$\eta_c = \left(\frac{l_1}{\sqrt{3} \cdot R}\right)^2 \tag{5.4}$$

$$\eta_c' = \left(\frac{t_2}{\sqrt{3} \cdot R}\right)^2 \tag{5.5}$$

The solution for the square of the resonance frequency for the rectangular and the circular mirror can be written as:

$$\omega^2 = \frac{6 \cdot EI}{m_2 \cdot l_1^3 \cdot (1 + \eta')} \cdot (1 + \eta' + \eta - \sqrt{1 + \eta'^2 + 16\eta^2 + 4\eta + 4\eta\eta' + 2\eta'}) \tag{5.6}$$

When Eqn. 5.6 is plotted (Fig. 5.4) and compared with experimental results, the deviation between measurement and calculation is about 60% as already observed for the lateral one-dimensional mirror. However, the relative variations, when geometrical properties changed, remain the same in both cases. The differences between measurement and calculation can be explained by the neglected curvature and the uncertain Young's moduli. The clamping conditions and with them the definition of the beam length l_1 may certainly also be a cause for divergences. Besides these reasons, one can add the fact that the torsion of the long beam (l_1') and its influence on the deflection of the small beam is not taken into account.

Variation of the length of the long beam (l_1') does not influence notably the natural frequency as observed in Fig. 5.4 (measurements). As expected, the width or the radius of the mirror and the deflection beam length are of great interest and determinant in the result of the natural frequency.

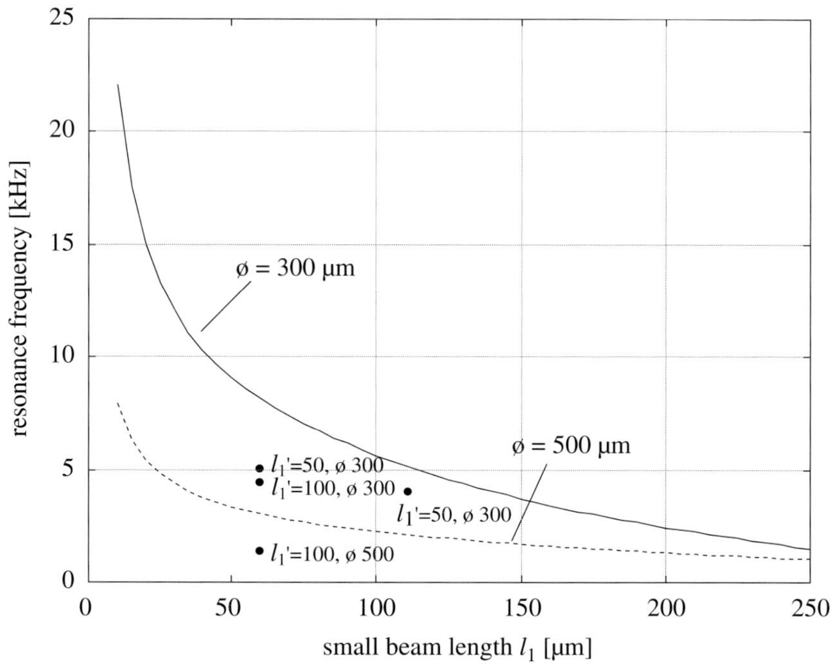

Fig. 5.4 *Calculated (lines) and measured (•) fundamental natural frequencies for two different mirror sizes (thickness of the mirror is 6.5 µm) The graph plots the resonance frequency versus small beam length (l_1) for different mirror diameters (ø) and large beam length (l_1').*

5.2.2 Simulations

Table 5.1 Simulated and measured resonance frequencies for the type 1 micro-scanner, mirror thickness of 5 μm.

	1st resonance mode [Hz]	2nd resonance mode [Hz]	3rd resonance mode [Hz]	4th resonance mode [Hz]
simulated frequency	294	330	1410	2350
experimental frequency	235	258	1345	-

Taking into account large deflection theory and initial curvature of the device, numerical modelling has been carried out using the finite element method with the commercially available package ANSYS 5.5. To take into account the effects of the different material layers, the "L"-shaped beam and the mirror were simulated using multilayered structural shell elements. As for the one dimensional model, the initial curvature is imposed in an arbitrary way and is not given by stress causes in the beams. Using this model, modal analyses of the scanner have been performed.

Table 5.1 lists the simulated and measured frequencies for the first three modes of the type 1, 2D-microscanner. The simulated mode shapes of the device correspond to the observed experimental movements. The experimentally measured natural frequencies are up to 25% lower than the simulated values. The deviations can be explained by inaccuracies in modelling the geometry of the device including the curvature of the beam (difficult to define) and, as for the 1D-microscanner, differences in the assumed and actual material properties (these properties can change with temperature and this effect is not included in the simulation).

The influence of the initial curvature was simulated in Fig. 5.5 for the different modes of the scanner type 1. It can be seen that the frequencies increase when the curl angle decreases, what is confirmed with the analytical model, which neglects the curvature. Simulations also showed that the mode shapes change for certain configurations. Note the good correlation between experimental and simulated values.

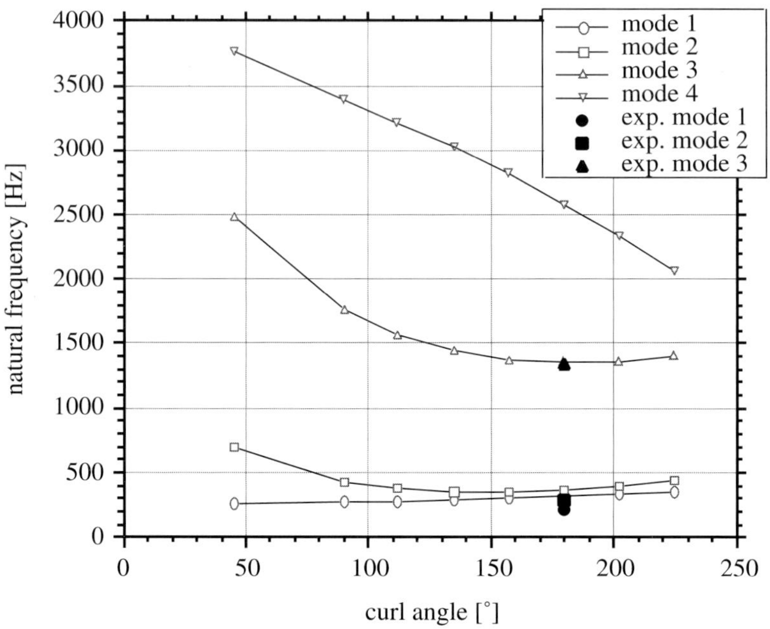

Fig. 5.5 Simulated natural frequencies for given initial curvature position of the mirror type 1. The curl angle is the sum of the position angle of the short beam and the position of the long beam. Experimental points are in black.

5.3 DYNAMIC MEASUREMENTS

5.3.1 2D-microscanner type 1

The typical scanning amplitude vs. driving frequency has been determined by measuring the deflection of a reflected laser beam on a target screen. AC heating with the appropriate frequency can excite several mechanical resonances (Fig. 5.6). For a typical mirror size of 500 by 300 μm^2, with a 300 μm by 120 μm long "L"-beam and 100 μm in width (type 1), the first resonance frequency of the motion around the *y*-axis is at $\omega_1 = 235$ Hz. The first mode of torsional motion, the raster scanning (the scanner twists around the *x*-axis, Fig. 5.2 a) is observed at $\omega_3 = 1345$ Hz. The first peak at 117 Hz is an artifact of the thermal actuation.

Fig. 5.7 shows the relationship between the driving power and the optical scanning amplitude for two driving frequencies. For the non-resonant (slow

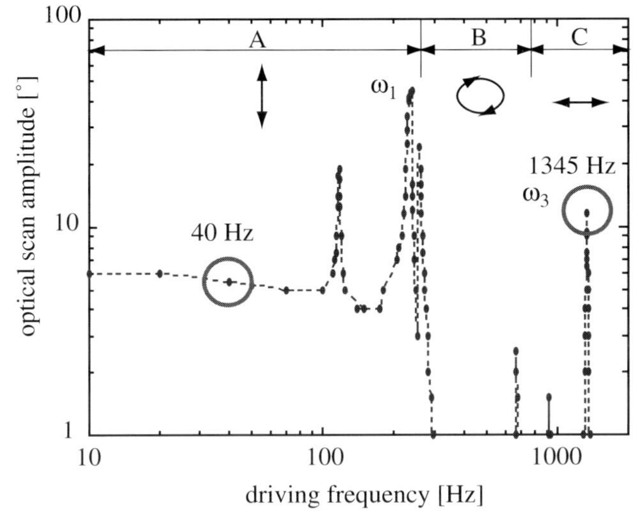

Fig. 5.6 Single directional amplitude frequency response of the type 1 scanner for positive sine excitation at 2 mW. A: y-axis movement, B: xy mode coupling region, C: x-axis resonance. The two circles (at 40 Hz and 1345 Hz) show the frequencies that are combined in the composite signal for image projection.

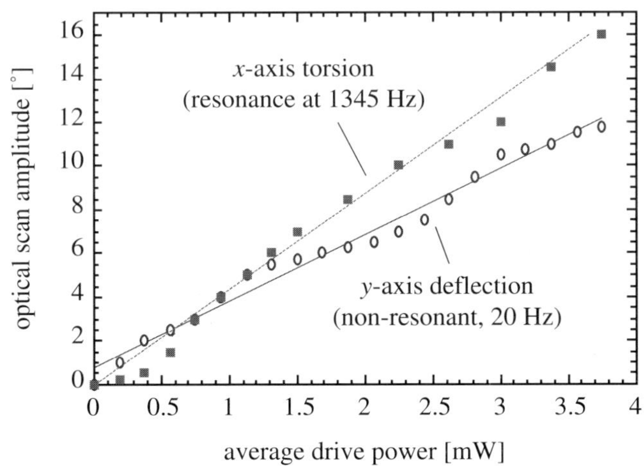

Fig. 5.7 Optical scan angle vs. drive power of the type 1 scanner.

motion) and the resonant mode, typical optical scanning amplitude of 10° is measured for an average drive power of 3 mW.

5.3.2 2D-microscanner type 2

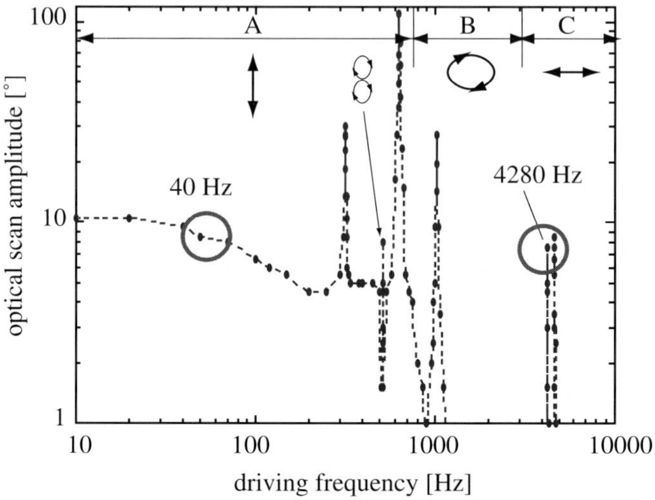

Fig. 5.8 Single directional amplitude frequency response of the type 2 scanner for positive sine excitation at 6 mW. A: y-axis movement, B: xy mode coupling region, C: x-axis resonance. The two circles (at 40 Hz and 4280 Hz) show the frequencies that are combined in the composite signal for image projection.

To improve the raster scanning frequency, stiffer and shorter beams were designed. The new microscanner has a 300 µm diameter round mirror with a 50 µm by 100 µm long "L" beam and 100 µm in width (Fig. 5.2 b). Similar to the frequency response of microscanner type 1, three modes can also be distinguished (Fig. 5.8). The fundamental resonance frequency of the motion around the y-axis is at 633 Hz. A coupled mode (which seems to be the second mode) appears at 1023 Hz. The raster scanning (the scanner twists around the x-axis) is observed at 4280 Hz. For the non-resonant and the torsion motion, a typical optical scanning amplitude of 7° is measured for an average drive power of 6 mW. The raster scanning frequency of 4280 Hz allows, if combined with 40 Hz frame scanning signal, the definition of more than 100 lines (section 5.5.2).

5.3.3 Summary

The design and modelling as well as the experimental results of the two-dimensional scanners have been presented and compared. Good correlation could be found between simulated and measured values. A qualitative estimation of any change in the design can be easily predicted.

Type 1 of the 2D microscanner is capable of 15° optical scan amplitudes in two orthogonal directions for a typical power consumption of 5 mW. With a modulated light source, images can be displayed. The type 2 micro-device shows promising performance improvements for the raster scanning application. The combination of a high resonant raster frequency at 4280 Hz with a 40 Hz image refreshing frequency allows 100 lines of definition.

5.4 MICROPROJECTOR APPLICATION

The presented two-dimensional microscanner with an "L" shaped cantilever beam (Fig. 5.9) is capable of two orthogonal angular motions. The movements are simultaneously excited at non-resonance (the frame scanning) and at resonance frequency (raster scanning). This section evaluates the utilization of the microscanner in display applications, precisely in raster-scanning laser displays.

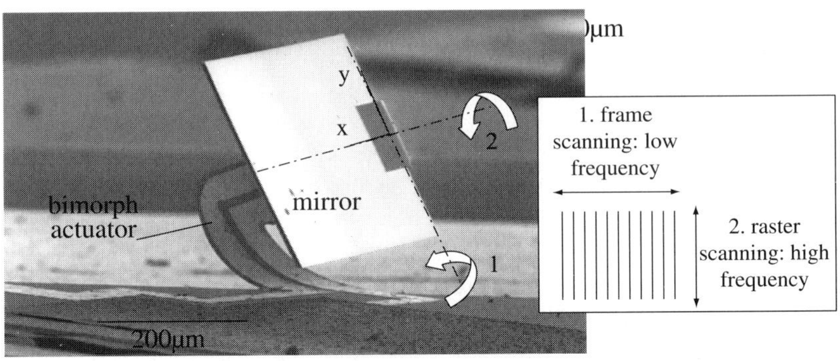

Fig. 5.9 Mirror Type 1: mirror size of 500 by 300 µm² rectangular mirror, with a 300 µm by 120 µm long "L"-beam.

5.4.1 Scanner requirements

ΘD is the product of mechanical scan angle and mirror size, which is shown as the defining parameter for resolution in scanned displays [117, 120, 163]. Likewise, the display refresh rate requirement translates into frame (in our case horizontal, see Fig. 5.9) and raster (in our case, vertical) scanner frequencies (f_f and f_r). Both the ΘD-product and f_r are very important figures of merit for scanner performance comparison.

Table 5.2 Video performance parameters for scanned displays [120].

Video performance level (pixel x line)	ΘD for 635 nm [degrees·mm]	Raster scan frequency f_r @ 60 Hz [Hz]
QVGA (320 x 240)	3.75	7800
VGA (640 x 480)	7.5	15750
SVGA1 (800 x 600)	9.37	19200
SVGA2 (800 x 600)	6.25	19200
SXGA (1280 x 1024)	15	32000

A designer can use this ΘD data to choose a mirror size and a scan angle that will best optimize the size and power requirements for a given application. In general, the mirror size D is chosen based on overall size, dynamic mirror flatness and maximum allowable scan angle due to optical constraints. The actuator must be sensitive and stiff enough to allow to scan to the required angle at the required frequency. There are also drive circuit requirements and cost trade-offs which must be considered.

5.4.2 Pixel resolution

The general scan equation [117] for vertical and horizontal resolution of a scanning display can be written as:

$$N_v = \frac{4\Theta_{max} D K_{sp} K_{os} \cdot \cos(\alpha)}{K_t a \lambda} \tag{5.7}$$

$$N_h = \frac{4\Theta_{max} D K_{sp} K_{os} K_{ub} \cdot \cos(\alpha) \cdot n}{K_t a \lambda} \tag{5.8}$$

$$f_r = \frac{f_f N_h}{K_{ub} K_{os} n} \tag{5.9}$$

where

- N = number of pixels, resolution
- f_r = raster scanner frequency (Hz)
- f_f = frame rate or refresh rate for the image (typically 50 or 60 Hz)
- D = scan mirror size
- Θ_{max} = maximum mechanical scan amplitude
- α = feed beam angle, which also affects pixel size resolution. When α is increasing from vertical ($\alpha = 0$), the projected area of spot on the mirror increases as a cosine function.
- a = mirror shape factor (a = 1 for rectangular mirror, a = 1.22 for circular mirror)
- λ = wavelength
- K_{sp} = spot size to pixel size ratio: A K_{sp} of 1 corresponds to Full-Width-Half Maximum (FWHM) overlap of adjacent pixels and is considered the optimal trade-off between raster modulation and Modulation Transfer Function (MTF). Larger K_{sp} values correspond to more overlap, which will lower contrast and MTF.
- K_{os} = overscan constants (ð1): The speed variations resulting from sinusoidal motion of resonant scanners produces pixel size and brightness variations across the scan line. This can be easily corrected electronically. Typical value are between 0.7 and 1 [117].
- K_t = beam clipping constant: is function of the amount of beam clipping at the limiting system aperture (mirror). Typical value: between 1.04 and 1.5 [117, 164].
- K_{ub} = 1 for unidirectional and 2 for bidirectional scanning: bidirectional scanning refers to writing a new line or column of data in both scan directions as opposed to writing data only during the forward sweep. Bidirectional scanning increases the light source modulation by a factor of 2 and reduces the required raster scanning frequency by a factor of 2. However, it requires precise control of the phase between forward and backward scan sweeps and can produce non uniform line-to-line or column-to-column spacing across the scan.
- n = number of feed beams: using multiple beams to scan multiple columns increases the resolution by increasing the number of lines or columns written during a simple sweep of the scan mirror. f_r is inversely proportional to the number of beams. The latter will depend on factors such as complexity of the overall systems, cost and optical alignement considerations.

Several other factors can affect image quality. For instance, the non-linearity of the frame scan should be less than 0.1% across the scan line. The high frequency scanner position and its frequency need to be determined very accurately and have to be monitored continuously.

Optical and mechanical design of the scanner housing imposes additional constraints. Manufacturing and alignement tolerances become very tight for small scan mirrors. Small mirrors require large scan angles, and large scan angles result in focus errors and non-flat image plane (i.e. field curvature). Field curvature can be corrected optically, but for low cost and low weight display applications, there is no room for field correction lenses.

5.4.3 Static and dynamic mirror deformations

Both the static and the dynamic flatness of the mirror are important for image quality in any optical scanning application. In Chapter 3, we worked on the flatness and the rigidity of the mirror, proposing annealings and reinforced plates, which resulted in flatness improvements and in a curvature radius of 12 mm for a mirror of 300 μm in diameter. This corresponds to a maximal static mirror deformation of about 1.8 μm.

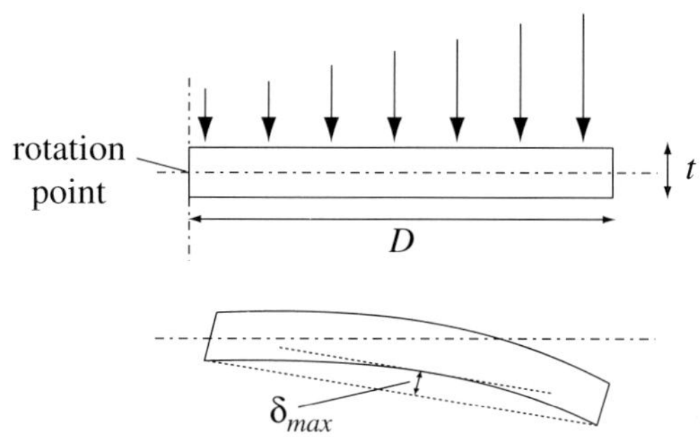

Fig. 5.10 Graphical representation of a dynamic mirror deformation. The bending of the mirror plate is due to the inertial moment.

As illustrated in Fig. 5.10, high acceleration forces during mechanical deflection of the mirror result in the bending of the plate. Maximum deformation occurs at the extremities of the scan. For rectangular mirrors, the following formula [117, 119] can be used to compute maximal mirror deformation δ_{max}

(Eqn. 5.10), which is defined as the deviation from linearity as shown in Fig. 5.10.

$$\delta_{max} = 6.94 \cdot \frac{\rho \cdot f^2 \cdot D^5 \cdot \Theta_{max}}{E \cdot t^2} \tag{5.10}$$

where

- ρ = material density
- f = driving frequency
- D = mirror length
- Θ_{max} = maximum mechanical scan amplitude
- E = modulus of elasticity (Young's modulus)
- t = mirror thickness

For the kind of aberration the mirror introduces, the maximum mechanical deformation (δ_{max}) should not exceed $\lambda / 10$ ($\lambda / 5$ optical path difference) of the shortest system wavelength. This will permit the spot to remain diffraction limited across the scan line. Note that δ_{max} is proportional to $D^5\Theta$. Due to the high sensitivity to changes in D, if the system performance is deformation limited, significant gains can be attained by slightly reducing mirror size.

5.4.4 Experimental results

Two-dimensional laser beam display

A simple setup has been built for image displaying (Fig. 5.11). The light is emitted by a laser diode (1 mW, 635 nm) and focused on the mirror of the microscanner. The computer controls simultaneously the modulation of the laser and the movement of the scanner. The laser beam is projected on a small screen.

Two-dimensional laser beam scanning is demonstrated where the modulation of the laser diode provides image projection as shown in Fig. 5.13 a and b. The displayed picture corresponds to an optical scanning angle of 10°. In this experiment, the square root of the sum of two positive sine voltages was imposed (Fig. 5.12); one at 40 Hz, exciting the non-resonant motion (frame scanning), the other at 1345 Hz, driving the resonance torsional motion (raster scanning). This was done in order to achieve a sinusoidal variation of the injected power (because power goes with the square of the voltage), and thus a sinusoidal variation of the temperature in the actuator.

Fig. 5.13 represents the scanning frames in one scanning direction (the pulsed laser is only illuminating the mirror during the forward scanning frame), what we called unidirectional scan in §5.3.2, showing 15 scanned columns.

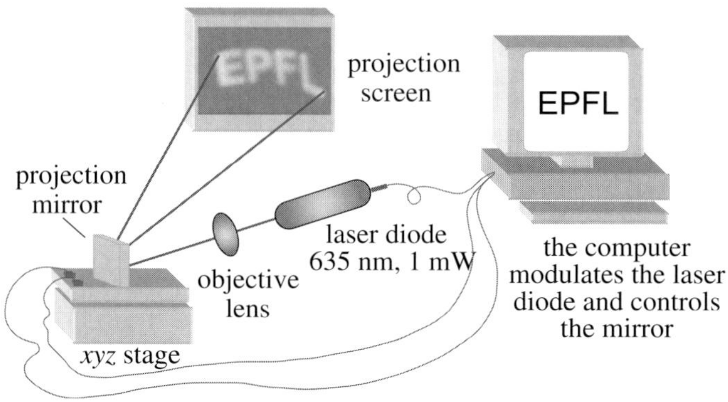

Fig. 5.11 Setup for laser scanning display.

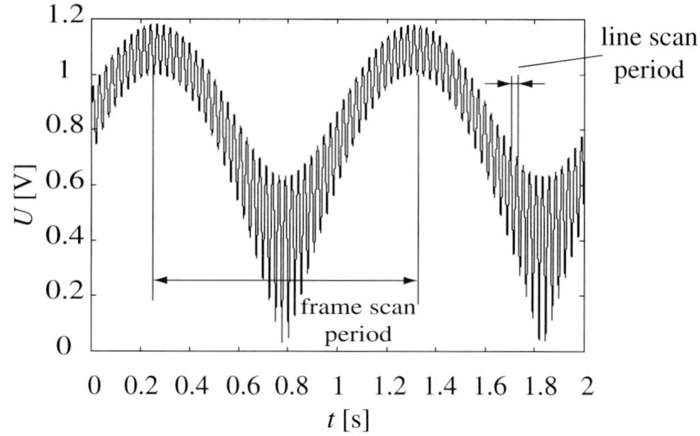

$$U=\sqrt{(B+A_1 \sin(\omega_1 t)+A_2 \sin(\omega_2 t))}$$

Fig. 5.12 Driving voltage applied to the device.

The image sent by the computer and the display projected on the screen are compared in Fig. 5.13. Note that in these projections the frame scanning is from left-to-right (low frequency: 40 Hz) and the high frequency scanning from top-to-bottom (1345 Hz).

By displaying a 3 x 5 dots array (Fig. 5.13 b), it is possible to quantify the performance of the microscanner. One can see some distortions in the image. The projected frame is not square; it forms an arc. This can be explained by the fact that the two movements are not totally independent one from the other. In particular, the temperature variation in the entire beam caused by the

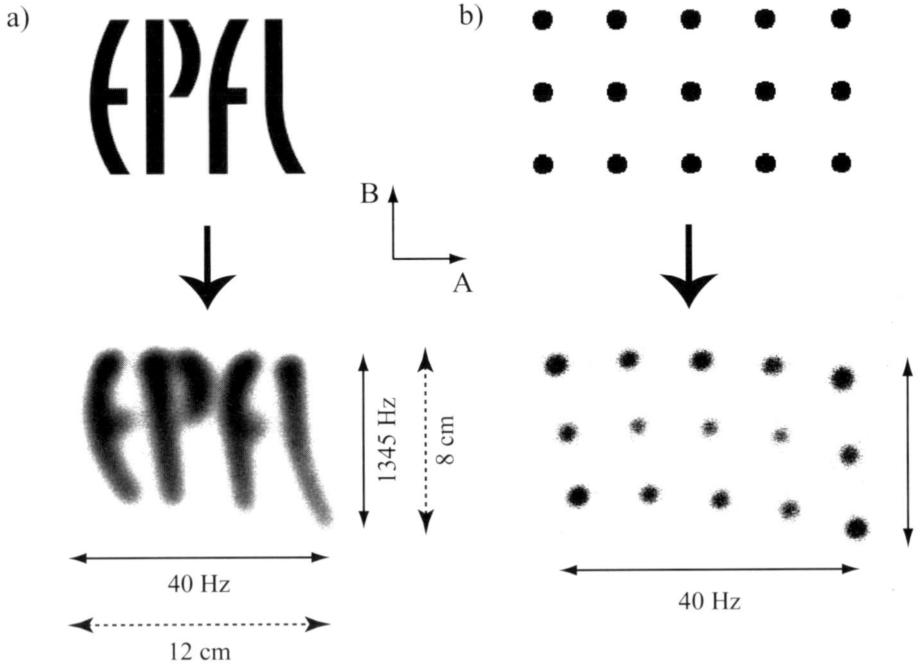

Fig. 5.13 a) Projection of the logo "EPFL" (Ecole Polytechnique Fédérale de Lausanne), b) Image of a 3 x 5 square array of dots.

non-resonant frame scanning (at 40 Hz: from left-to-right) creates an offset of the vertical raster scanning position (direction B, from top-to-bottom). The average temperature is the highest on the right end of the picture (direction A).

The dot size also varies across the image, indicating variations in optical resolution across the field of view. This can be explained by the curvature of the mirror. The static mirror curvature has not been optically corrected in the set-up. We observe a poor image resolution in the edges of the display.

Fig. 5.14 shows the resonance frequency of torsional motion as a function of the DC power in the actuator. For this measurement, a variable DC bias voltage is applied to the bimorph (the amplitude of the sine is fixed). It is observed that increasing the power in the beam, thus raising the average temperature in the actuator slightly increases the resonance frequency. This illustrates that the resonance frequency can change with temperature and thus denotes the importance of an accurate control of the frequency.

Compensation of this effect needs some dynamic adaptation of the signal driving the torsional resonance motion. Detecting the resonance frequency

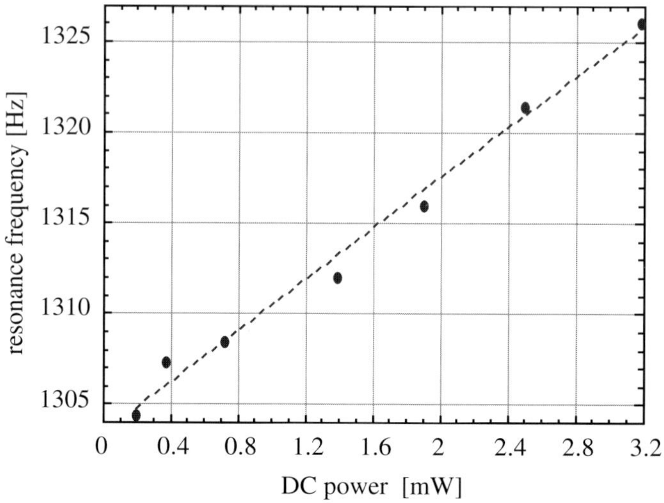

Fig. 5.14 Frequency shift vs. drive power of the 2D scanner. The average AC drive power is 0.3 mW.

electronically (strain gauge effect) and adapting simultaneously the driving frequency of the high frequency scanning mode could solve the problem.

Discussing resolution issues

The resonance frequency used for the raster scanning limits the display resolution by restricting the number of columns that can be scanned during the image refresh period. The demonstration micro-device has a raster scanning frequency of 1345 Hz that allows, if combined with 40 Hz frame scanning signal, the definition of more than 30 columns (if the bidirectional scanning is used). To increase the number of columns, the resonant frequency of the raster scanning (high torsional frequency mode) has to be increased.

A new design has been made resulting in the microscanner type 2. An effort has been done in working on the planarity of the mirrors and stiffening them. The dimensions of the beam has been modified as well. The raster scanning frequency of 4280 Hz allows, if combined with 40 Hz frame scanning signal, the definition of 101 columns (N_h) (Eqn. 5.9) for the values of Table 5.3. From Eqn. 5.8, a ΘD product of 2.1°·mm can be calculated and a horizontal number of pixel N_v of 129 deduced from Eqn. 5.7. Summarizing, a resolution of 101 x 129 pixels is obtained.

The quantity of light increases at the border of the frame. This can be explained by the sinusoidal speed variation (therefore moving the slowest at the

Table 5.3 Parameters for the calculation of type 1 microscanner resolution.

$K_{os\text{-}h}$	$K_{os\text{-}v}$	K_{sp}	K_t	K_{ub}	λ [nm]
0.95	0.8	1	1.5	1	635

ends and the quickest in the middle). We compensated this effect by an adapted intensity modulation of the laser. This homogenizes the light quantity during the scanning. The vertical overscan factor $K_{os\text{-}v}$ accounts for the unused portion of the scan. Similarly, the horizontal overscan factor $K_{os\text{-}h}$ takes into account the retrace time of the non-resonant scan.

K_{sp} and K_t (corresponding to a full Gaussian beam, with no clipping) are arbitrary values as the spot dimension and the quality of the reflected beam is highly dependent from mirror quality. $K_{ub} = 1$ as the scanning is unidirectional and $n = 1$, as we use only one feed beam. The incident angle α is considered to be equal to zero (the incident beam is perpendicular to the mirror).

Plotting ΘD versus scan frequency illustrates the performance of different scanners, as compared to each other and the video standards. Fig. 5.15 shows a graph of a number of published results. To meet a given video standard, a scanned display must fall above and to the right of the indicated lines.

This graph points out the high performance requirements for useful videos. Note that this graph does not speak of static and dynamic flatness requirements, which are additional constraints to the system. Video requirements on mechanical scanners are demanding, since they need high scan angles and high frequencies. Large mirror sizes relieve the angle requirement, but become difficult due to flatness and other considerations.

Some of these MEMS mirrors have other target applications with sometimes lower requirements as those for the video display. Our microscanner shows quite low video performances. But, fabrication costs are low in our case (Chapter 3), compared to the other mentioned MEMS mirrors. Indeed, the latter are usually based on a 2 to 5 times higher mask process than the presented microscanner.

Discussing static and dynamic mirror flatness

For the case of our microscanner (Table 5.4), the values of δ_{max} (Eqn. 5.10) was calculated to be better than $\lambda/10$ in the visible wavelength range. Indeed, δ_{max} varies between 4 to 30 nm for maximal mechanical angles between 5° to 30°, frequencies from 300 to 5000 Hz and mirrors from 300 to 500 µm. If planning higher frequencies up to 20 kHz, especially in the image projection applications, the dynamic mirror deformation begins to be critical.

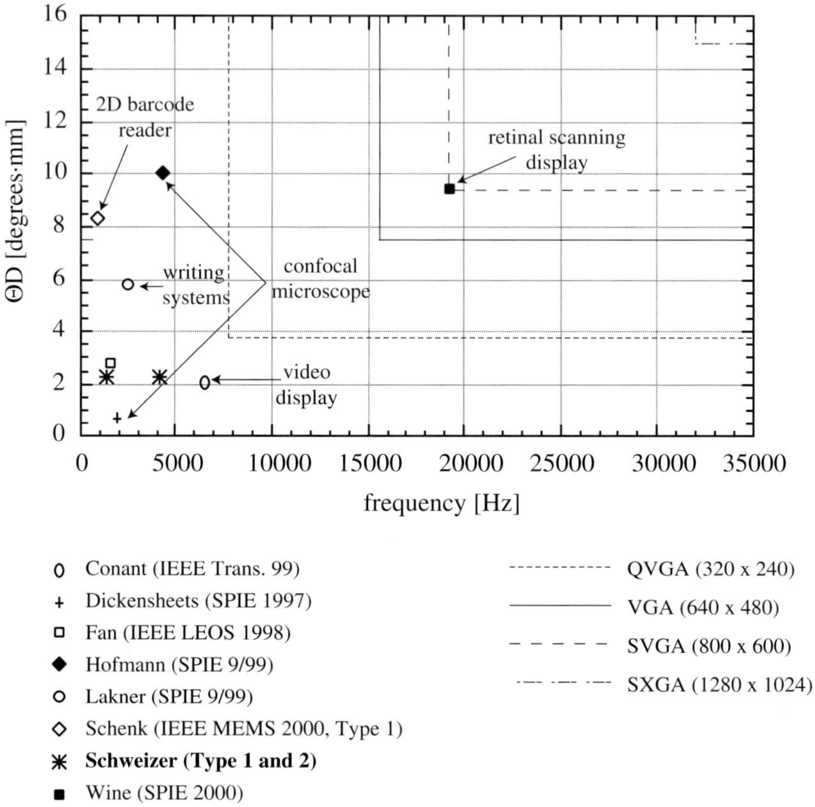

Fig. 5.15 Comparison of various MEMS scanner mechanisms. The target application is indicated (if mentioned in the publications). Adapted from Wine [120].

In our case, the biggest issue concerning mirror flatness is still the static curvature and the fact that the axis of rotation is a function of the mirror position.

Table 5.4 Typical mirror plate values.

Designation	Value
ρ_{Cu}	$8.95\ 10^3$ kg/m^3
E_{Cu}	120 GPa
t	6 μm

5.5 CONCLUSIONS

A new two-dimensional microscanner was designed, fabricated and characterized. Simultaneous thermomechanical excitation of a combined resonant and non-resonant movement of an "L"-shaped beam has been demonstrated. It is capable of maximal mechanical scanning angles of 7° for 5 mW in the two scanning directions. The raster scanning frequency varies from 1000 Hz to 5000 Hz depending on the geometry. This was achieved by adjusting geometrical parameters such as mirror dimension and thickness, beam length and thickness and finally beam configuration. Analytical models and simulations for the determination of the raster scanning frequency were presented and compared to the experimental results.

With a modulated light source, successful image displaying was demonstrated. The resolution of the device remains quite low, however. In particular, the raster scanning frequency and the mirror static flatness are issues to be improved. High performance displays impose very stringent requirements on the scanners. In summary, three most important figures of merit for scanner evaluation can be stated as scan angle - mirror size product (ΘD-product), scanner frequency, and static and dynamic surface flatness for raster scanning applications. Until now, only few MEMS mirrors meet these specifications.

Doors are open to other applications such as two-dimensional barcode readers where the required scanning frequencies are lower than for video displays.

6 ADVANCED OPTICAL FILTERS OF POROUS SILICON

The important thing in science is not so much to obtain new facts as to discover new ways of thinking about them.

Sir William Bragg

6.1 PRINCIPLE

Porous silicon is fabricated by electrochemical etching of monocrystalline silicon in hydrofluoric acid (HF). A multilayer dielectric interference filter of porous silicon is based on the effect that the index of refraction depends on the degree of porosity, which can be chosen individually for every layer by the fabrication process. The porosity, i.e. the ratio of void volume to total volume in the layer, depends on the current density during the layer formation by electrochemical etching. To fabricate an optical filter the etch current is varied over time according to the desired refractive index profile of the layer stack. With this technology, optical filters of porous silicon can be produced within minutes in a single process step that generates a modulated density (Fig. 6.1).

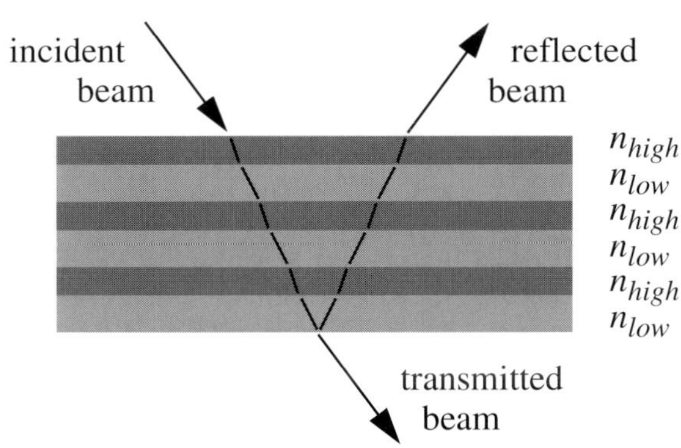

Fig. 6.1 Optical interference filter made of alternating porous silicon layers with different porosities resulting in different refractive indices.

6.2 HISTORY OF POROUS SILICON

As often in the history of science, when porous silicon was discovered, it was thought as something unwanted rather than useful. Uhlir [165] wanted to clean and shape silicon by electrochemical etching in hydrofluoric acid solutions, when he noticed in 1955 that a certain "threshold current density" is needed for smooth etching. "Lower current densities give black or red surfaces ..." he noticed. He did not manipulate in the laboratory himself, "Most of the experiments mentioned in this paper were carried out by my wife, Ingeborg", he acknowledges. Turner, also working at the Bell Telephone Laboratories, described the film as brown and remarked already in 1958 that "several orders of interference colors can be seen as the film thickens" [166]. Further he found that "Pieces of the film react with explosive violence when put in contact with a strong oxidizing agent such as nitric acid". The nature of this film was first thought to be amorphous silicon [167], later silicon hydroxides [168].

It took until 1972 when Theunissen reported that pores formed in the film [169]. For a long time, porous silicon did not attract much attention, until in 1990 Canham found that nanoporous silicon is photoluminescent and can emit visible light [170]. Together with the discovery of electroluminescence in porous silicon, this launched an avalanche of research, resulting in more than 3500 papers about porous silicon in the year 2000 [171]. Cullis *et al.* gave a good review about the properties of porous silicon in 1997 [172] and several books have been published on that topic [173-174]. However, the hope for optoelectronic devices like LEDs of porous silicon was disappointed, mainly because of poor quantum efficiency.

The knowledge gained from this basic research on porous silicon can on the other hand be useful for other domains. All basic material properties have been measured as well as their dependence on fabrication parameters. It is now time to find new applications for this extraordinary material. One of the first commercial products using porous silicon was a high-density capacitor presented by Lehmann at Siemens [175], making use of the huge internal surface of macropores. Another product are SOI-Epi wafers fabricated at Canon Inc. by bonding and selective etch-back of porous silicon on a carrier wafer [176, 138 (chapter 12.9)].

6.3 FABRICATION OF POROUS SILICON

Porous Si forms by electrochemical etching of *p* or *n* doped monocrystalline Si in hydrofluoric acid (HF) at current densities below a so called critical current density. Fig. 6.2 shows the top view of a porosified wafer. Although the exact

Fig. 6.2 SEM photograph of a porous silicon surface, fabricated from p^+ doped [100] silicon (0.012 Ωcm), current density 150 mA/cm^2.

explanation of the formation process is still object of discussions [177], the overall reaction, starting with a hydrogen terminated Si surface and two positive charge carriers h^+ is

$$=SiH_2 + 2\,F^- + 2\,h^+ \rightarrow SiF_2 + H_2 \text{ (2-valent dissolution)} \tag{6.1}$$

$$SiF_2 + 4\,HF \rightarrow 2\,H^+ + SiF_6^{2-} + H_2 \text{ (chemical step)}$$

As a surprising effect for an anodic reaction, hydrogen is formed at the surface [173].

Besides Si, other semiconductors can be porosified, like GaAs that also shows photoluminescence [178] as well as SiC [179]. For the application in this work, only Si was investigated, because it is compatible with micromechanics, it is by far the more economic material and most of the data is available for Si.

6.3.1 Single etch cell

The most simple apparatus for the fabrication of porous Si is an electrochemical single cell, Fig. 6.3, left side. It is filled with a HF electrolyte solution. The Si wafer is mounted in a wafer holder and immersed in the cell, facing a counterelectrode of platinum. To achieve an ohmic contact to Si, it has to be heavily doped (e.g. p^+) on the backside, or a metal layer has to be deposited on the backside as contact and driven in. A constant current is passed through the cell to porosify or electropolish the wafer.

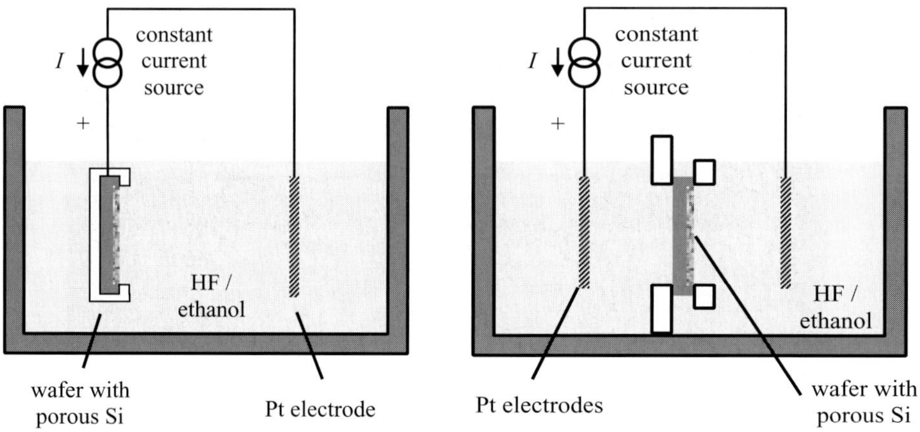

Fig. 6.3 Electrochemical cells for porosification and electropolishing of Si. Left: single cell, right: double cell with wet backside contact.

6.3.2 Double etch cell

For the work presented here, an electrochemical double cell was used, Fig. 6.3, right side. It contains two platinum electrodes, one in each half of the cell, with the wafer in the middle, completely separating the two parts. The current is forced to pass through the wafer, etching its anodic side. A p^+ doped wafer makes an ohmic wet backside contact with the electrolyte. The uniformity of this process is better than with single cells, because it avoids lateral currents at the backside, as they appear using a dry contact. The etch bath can be stirred with a magnetic stirrer bar to improve homogeneity. A standard electrolyte consisting of HF (50% wt.) and ethanol (C_2H_5OH) in a volume ratio of 1 : 1 was used.

Security is an issue when handling HF, it causes severe burns after some latency time [180]. The advantage of the installation used is that the electrolyte can stay in the cell for a long time and one needs not to touch HF for processing. The wafer can be immersed and taken out at a handle and fixed with a twist-lock mechanism to press it against the cell separation wall (Fig. 6.4).

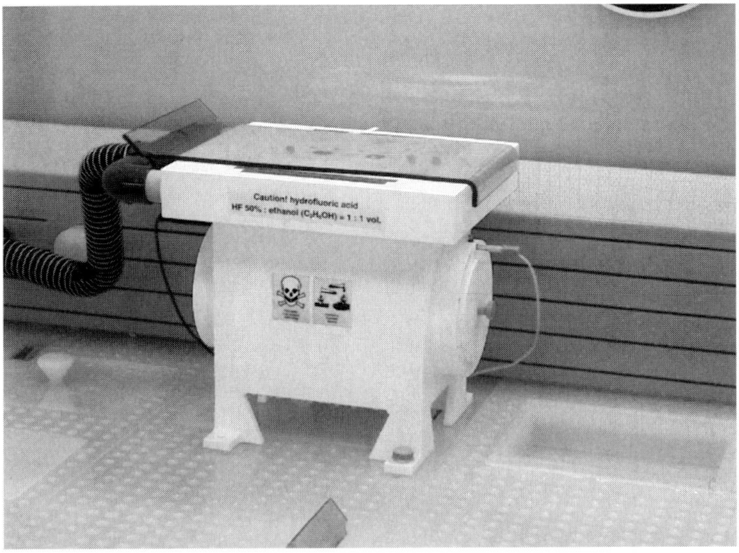

Fig. 6.4 Electrochemical double etch cell for 100 mm wafers installed on a wet etch bench in a cleanroom. The vapor extraction can be seen on the left.

6.4 PARAMETERS DETERMINING THE STRUCTURE OF POROUS SILICON

6.4.1 Why porous silicon is porous

There is a wide variety of morphologies of porous Si. The reason for the formation of pores is a strong self-roughening effect of the reaction at the surface. When the Si surface is some atoms higher at one location, the semiconductor band structure will be changed in such a way, that the material becomes less conductive, the probability to be etched gets even lower and a peak remains. New charge carriers from the substrate arrive first at the grooves and induce etching there. Once a well is formed, etching continues at its bottom to make it a pore (Fig. 6.5). The etch process is charge carrier limited, not material limited.

The resulting structure is a sponge-like labyrinth of interconnected pores. The ratio of void volume in porous Si over total volume is defined as porosity. It can be determined by precision gravimetry or measuring the refractive index. The dependence of the properties on fabrication parameters will be discussed in the following section.

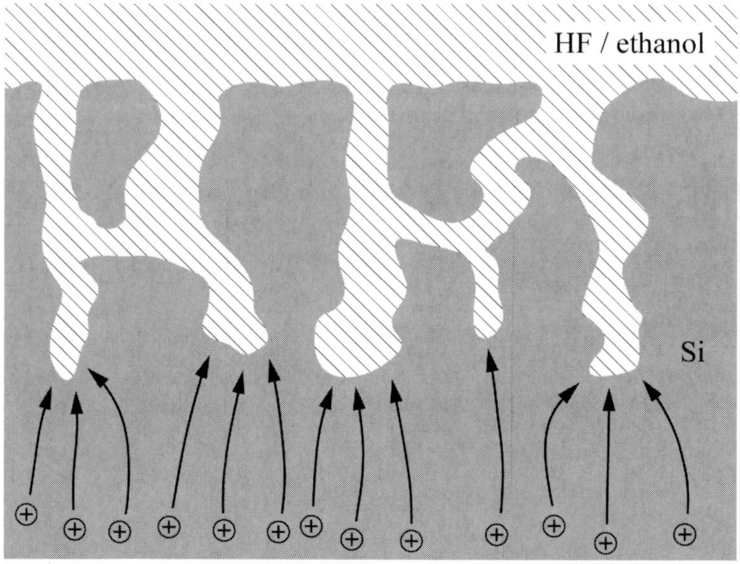

Fig. 6.5 *Formation principle of porous Si. The positive charge carriers (holes) etch Si where they get in contact with the electrolyte first, at the bottom of the pores, continuing the sponge morphology.*

6.4.2 Substrate doping

The pore size follows a distribution that depends mostly on the doping concentration [181]. It is used to classify porous Si according to the IUPAC (International Union of Pure and Applied Chemistry) recommendations in:

- microporous Si with pore dimensions < 2 nm,
  obtained from p^0 doped Si, (> 0.1 Ωcm),
  shows photoluminescence,
  has the highest internal surface of up to 5000 m^2/cm^3,
- mesoporous Si with 2 .. 50 nm pores,
  obtained from p^+ doped Si, (≤ 0.1 Ωcm),
  mechanically more stable,
  offers a wide range of porosities and refractive indices,
- macroporous Si with pores > 50 nm,
  obtained from *n* doped Si [182],
  single macropores can follow a given regular mask structure.

For the fabrication of optical filters in this work, mesoporous Si from substrates with a resistivity of 0.010 to 0.020 Ωcm was chosen. It is mechanically more stable than microporous Si and leads to more robust devices. A range of porosity from 25% to 75% can be achieved, which allows a broad range of indices of refraction for optical elements.

It was reported that p^+ substrates give much smoother interfaces in multi-layers than p^0 substrates, inducing less scattering of light [183]. On the other hand the absorption of infrared light by p^+ substrates is higher due to the high concentration of free charge carriers. This fact is of less importance when only thin films of porous Si are used that are detached from the substrate.

6.4.3 Illumination

Visible light on the wafer surface creates additional charge carriers. The interface Si / electrolyte works similar to the semiconductor / metal interface of a Schottky diode and rectifies the photocurrent, which adds to the applied etch current. In general it was observed that illumination inhibits porosification of *p* doped Si, whereas it supports the porosification of *n* doped Si. Also the morphology of porous Si changes under illumination [184]. It is difficult to control the photocurrent over depth because of absorption.

Porous Si can be patterned by a projected image on the substrate during fabrication to influence its formation [185]. High resolution surface gratings of porous Si have been obtained by illumination with holographic interference patterns [186]. For the application presented here illumination was not necessary.

6.4.4 Electrolyte concentration

Most of the experiments reported in literature use a mixture of concentrated (fuming) hydrofluoric acid of 50% in weight and ethanol (C_2H_5OH) in a volume ratio of 1 : 1. As the specific density of HF is 1.16 g/cm^3 compared to C_2H_5OH 0.789 g/cm^3, this does not correspond exactly to HF : H_2O : C_2H_5OH = 1 : 1 : 2 as sometimes referred in literature.

Ethanol lowers the surface tension to improve wetting of the porous surface. Isopropyl alcohol and other solvents have been used to investigate the influence of surface tension on the optical properties of porous Si [187]. The low viscosity of ethanol improves the exchange of electrolyte through the pores during etching.

Halimaoui [173] examined the influence of HF concentration and found that the porosity decreases with increasing HF concentration. The critical current density is higher for higher HF concentrations, and the range of porosities that can be fabricated is largest for high HF concentrations. For these reasons, the standard electrolyte described above was used.

6.4.5 Current densities

Most important for the fabrication of optical filters is the fact that the porosity depends on the etch current density. The increase of porosity with current density was found to be linear in first approximation [173]. A more precise analysis by Berger and Arens-Fischer revealed a dependence that was fitted best with a logarithmic function [188]. The following empirical formula was found for p^+ doped Si and standard electrolyte:

$$P(j) = P_0 + \gamma \cdot \ln\left(\frac{j}{\zeta} + 1\right) \qquad (6.2)$$

$$j(P) = \zeta \cdot \left(e^{\frac{P - P_0}{\gamma}} - 1\right)$$

with j being the current density, P the porosity and the constants $P_0 = 0.223$ (extrapolated porosity for 0 current), $\zeta = 34.7\ \mathrm{mA/cm^2}$ and $\gamma = 0.246$ obtained from curve fitting. This means that porosities from about 25% to 75% can be realized with current densities from almost 0 to 250 mA/cm^2, see Fig. 6.6. If the porosity exceeds 75%, the internal sponge structure becomes too fragile and the surface tension of water breaks the material during drying [189, 190].

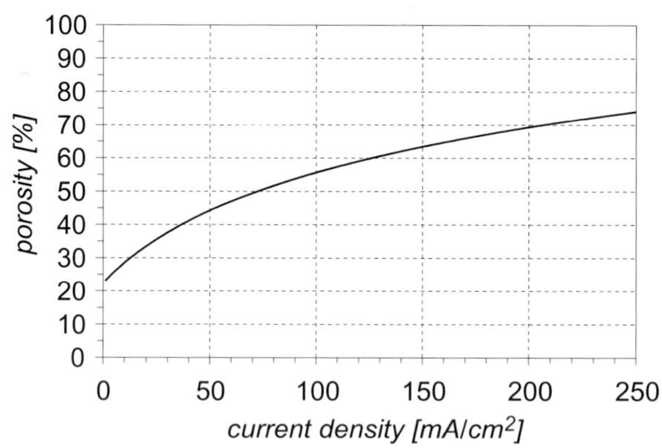

Fig. 6.6 The dependence of porosity on current density, shown for p⁺ doped Si and standard electrolyte, using the data from Ref. [188].

The voltage drop in the double cell is composed as follows:

$$U_{cell} = U_{Pt\ cathode} + U_{electrolyte} + U_{Si\ anode} + U_{wafer} + \dots \\ \dots + U_{Si\ cathode} + U_{electrolyte} + U_{Pt\ anode} \quad (6.3)$$

with $U_{electrolyte}$ being the voltage drop given by the electrolyte resistivity:

$$U_{electrolyte} = I \cdot \frac{d \cdot \rho_{HF/ethanol}}{A} \quad (6.4)$$

with I the total current, A the electrolyte cross section area, ρ the specific electrolyte resistivity of 1.85 Ωcm for standard etch solution, d the distance between the Pt electrodes and the Si wafer. The typical electrolyte resistance measured for the set-up used is 8.0 Ω. The wafer resistance is calculated the same way and can be neglected for p^+ substrates.

The equilibrium electrode potentials for a material X relative to the standard hydrogen electrode are given by the Nernst equation:

$$U_X = U_H^0(X) + \frac{RT}{z_r F} \ln \prod_i a_i^{v_i} \quad (6.5)$$

R is the gas constant of 8.31441 J / (mol K),
T the temperature in K,
z_r the reaction charge number (electrovalence), unitless,
F the Faraday constant of 9.648456 C / mol,
a_i the activity of the reaction partner i and
v_i its stoichiometric number.

The standard electrode potentials $U_H^0(X)$ for specific reactions can be found in tables [128], e.g.

$$SiF_6^{2-} \leftrightarrow Si + 6\,F^- + 4\,h^+ \qquad -1.24\ V \quad (6.6)$$

The activity is the product of concentration and activity constant and can be found in literature [192]. The electrolyte concentration on the porous surface of the wafer is more difficult to model than with ordinary electrodes, as the electrolyte exchange through the pores is diffusion limited.

For practical applications it is therefore more convenient to drive the etch cell in constant current mode than in constant voltage mode. Fig. 6.7 shows the voltage drop vs. current for the double cell used. The electrolyte resistance is a slope, superimposed to the electrochemical potentials.

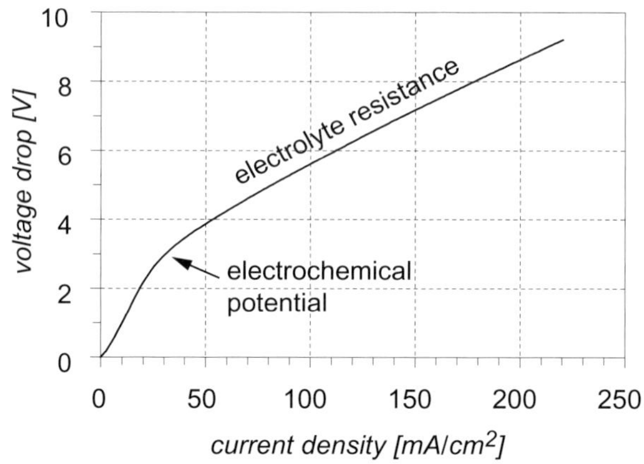

Fig. 6.7 Voltage drop versus etch current density in a double cell for p^+ Si (measured).

6.5 ELECTROPOLISHING

When the current density reaches a threshold value, the etch process is not charge carrier limited any more, but material limited, so that Si will be etched away entirely. This can be regarded as 100% porosity. The critical current density is about 500 mA/cm^2 for p^+ doped substrates in standard electrolyte. Electropolishing is isotropic electrochemical etching of Si.

It is remarkable that electropolishing also works under a layer of porous Si already formed and detaches the layer from the substrate [191]. This is possible because etching only takes place at the interface between porous Si and bulk Si, compare Fig. 6.5. The porous layer already built is almost an insulator and is etched only purely chemically by HF several orders of magnitude slower than electrochemically. Electropolishing is used for micromachining in chapter 4.

The chemistry of electropolishing is slightly different from porosification, it leads to a 4-valent reaction, compared to the 2-valent porosification [173]. More charge is needed to dissolve Si completely than to create pores of the same volume. In the transition region between porosification and electropolishing, a brittle brown film of highly porous Si forms that decomposes in dust during drying. This current region of transition has to be avoided if a clean optical surfaces is requested. The current should be switched within milliseconds from porosification to electropolishing.

When electropolishing under a layer of porous Si the critical current density is lower than at the surface because the electrolyte at the bottom of the porous layer is used and its HF concentration is lower. As soon as the porous layer is detached, fresh electrolyte can get to the substrate and the critical current density is back to its original value. When the current density is just at the limit to detach the porous layer, it can occur that it is not sufficient for continuous electropolishing of the substrate, but leaves residues of highly porous Si. This does not effect the optical quality on the bottom side of the porous layer.

6.6 POROUS SILICON AS SACRIFICIAL LAYER

Porous Si dissolves quickly in alkaline solutions like 1% KOH. In 40% KOH at 60°C, like it is used for bulk micromachining, porous Si dissolves in a violent reaction within seconds. Independent of the layer thickness, KOH penetrating the sponge structure of porous Si needs only to dissolve the walls of typically 20 nm thickness for mesoporous Si.

This effect has led to numerous applications for porous Si as sacrificial layer in micromachining. Lang reports free-standing bridges of polysilicon and metal [193, 194, 174 (chapter 19)]. He deposited the structures on porous Si, which was then dissolved as sacrificial layer. This technology was used to build a thermal sensor, the advantages over conventional bulk micromachining are discussed by Hedrich [195]. Also cantilevers of Si have been reported using porous Si as sacrificial layer [196], a good overview can be found in Ref. [197].

With a structured backside metallization to guide the etch current laterally it is even possible to underetch and porosify areas more than one millimeter wide under a membrane. After the dissolution of the porous layer the membrane served as thermally insulated hotplate [198].

In contrast to these applications, in this work porous Si is used as construction material for microsystems, especially for optical filters.

6.7 CALCULATION OF OPTICAL INTERFERENCE FILTERS

6.7.1 Principle

Berger *et al.* reported first in 1994 optical interference filters made of porous Si [199-202]. The current is changed during porosification to get layers of different porosities and refractive index. Fig. 6.8 shows a cleaved edge of such a multilayer structure. The method is patented [203]. Similar work was done almost at the same time by Vincent [204]. Various filter characteristics like band-reflectors, band transmission filters, edge filters, or broad band mirrors can be realized. Pavesi published a good review article about porous Si dielectric multilayers where he points out the capability of resonant luminescence of microcavities to control the emission spectrum [205].

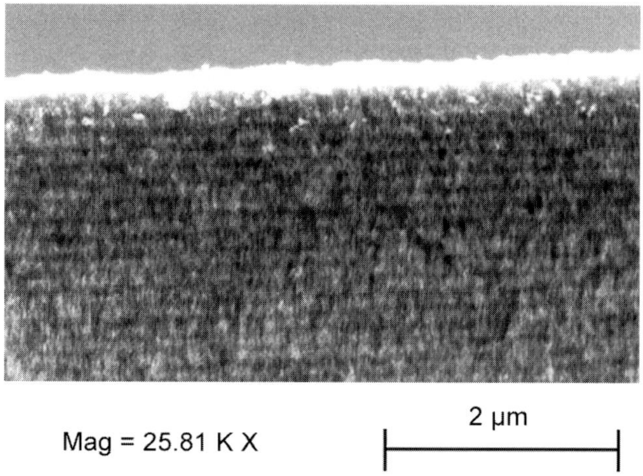

Fig. 6.8 SEM picture of a cleaved optical filter of porous Si. It consists of a stack of horizontal layers of Si with different porosities between 55% and 75%.

As the refractive index can be varied continuously within 1.5 and 2.5 for visible light (given by the fabrication range of porosity), this technology allows more degrees of freedom for the design than conventional layer deposition of different materials. The relation of porosity and refractive index will be treated in detail in section 6.8. First the necessary index profiles for different filter types shall be calculated.

6.7.2 General theory for simulation of optical multilayer filters

A good introduction to the theory of dielectric optical filters was written by Thelen [206]. The optical properties of a stack of N layers are calculated from the following input parameters:

n_l refractive index of layer l,
d_l geometrical thickness of layer l.

The optical thickness t_l is given by:

$$t_l = n_l \cdot d_l \qquad (6.7)$$

If the index profile is a gradient, it has to be sampled at discrete points.

A wave of wavelength λ that passes the layer l undergoes an optical phase shift ϕ_l

$$\phi_l = \frac{2\pi}{\lambda} n_l d_l \cdot \cos\alpha_l \tag{6.8}$$

where α_l is the angle of wave propagation in the film as determined from Snell's law

$$n_0 \sin\alpha_0 = n_l \sin\alpha_l \tag{6.9}$$

with α_0 the incidence angle to the normal of the filter surface and n_0 the refractive index of the surrounding medium.

The optical spectral transmission of an arbitrary layer stack is calculated using matrix theory. For every layer l of the stack, a characteristic matrix M_l is given by

$$M_l = \begin{bmatrix} \cos\phi_l & \dfrac{i\sin\phi_l}{n_l} \\ in_l\sin\phi_l & \cos\phi_l \end{bmatrix} \tag{6.10}$$

The incident wave, represented by its electric (E) and magnetic (H) field can be written as a vector to which the corresponding matrix product of the layer stack is applied, from bottom to top

$$\begin{bmatrix} E_1 \\ Z_0 H_1 \end{bmatrix} = M_1 M_2 \ldots M_N \begin{bmatrix} E_{N+1} \\ Z_0 H_{N+1} \end{bmatrix} \tag{6.11}$$

Z_0 is the impedance in free space $\sqrt{\mu_0/\varepsilon_0} = 377\Omega$.

Using this approach and applying some general electromagnetic relations, the transmission $T(\lambda)$ or reflectivity $R(\lambda)$ of an arbitrary multilayer interference filter are calculated. The theory can also be applied to absorbing materials, where transmission and reflectivity do not sum up to 1, which leads to complex indices of refraction.

The reverse problem of calculating the necessary index profile to obtain a given spectrum is not as straight forward. It can only be solved iteratively and challenges experts to worldwide competitions to find the best solution for a filter problem. An optimum solution does not exist generally. The mathematical effort in this field comprises e.g. genetic algorithms [207, 208].

Commercial programs for filter optimization are available and some of them have been tested, e.g. TFCalc. The calculation became very slow at 1000 layers and more, which is necessary to sample the gradient index profiles with adequate precision. The iterative optimization got stuck in suboptima; a system with several thousand free parameters is not evident to solve. Even with powerful genetic optimization algorithms a good set of start parameters is indispensable [208].

The most practical solution was to implement the simulation algorithm on a PASCAL compiler, where it runs one to two orders of magnitudes faster than a commercial program. It also allowed to integrate a procedure to generate user defined index profile functions automatically from a set of parameters. The profile was optimized iteratively by hand. Although the problem of filter simulation seems very intricate at a first glance, relations between the index profile and the resulting spectrum can be obtained analytically. After some experience, intuition grows for the refinement of the filter architecture, as will be shown in the following sections.

The simplest construction element is the quarter wavelength optical thickness (QWOT) layer, where

$$nd\cos\alpha = \lambda_0/4 \tag{6.12}$$

$$\phi = \pi/2 \tag{6.13}$$

leading to the matrix

$$M_{\lambda/4} = \begin{bmatrix} 0 & i/n \\ in & 0 \end{bmatrix} \tag{6.14}$$

and the half-wave layer with

$$nd\cos\alpha = \lambda_0/2 \tag{6.15}$$

$$\phi = \pi \tag{6.16}$$

$$M_{\lambda/2} = \begin{bmatrix} -1 & 0 \\ 0 & -1 \end{bmatrix} \tag{6.17}$$

6.7.3 Bragg band reflectors

An alternating periodic series of quarter-wave layers with high and low indices of refraction gives a Bragg band reflector, shown in Fig. 6.9. For a non-absorbing material, the filter ideally transmits the whole spectrum except the reflected band. The relative width of this first order diffraction is ideally:

$$\frac{\Delta\lambda}{\lambda} = \frac{1}{N} \tag{6.18}$$

where N is the number of periods of the stack, similar to a line grating with N lines. The band broadens additionally for high modulations of the refractive index. Reflectivities of more than 99% can be achieved, but the filter shows some imperfections compared to an ideal band reflector. The reflected band does not fall off steeply but contains sidebands. Higher harmonics appear in the reflected spectrum and it has a certain background, depending on the refractive indices used.

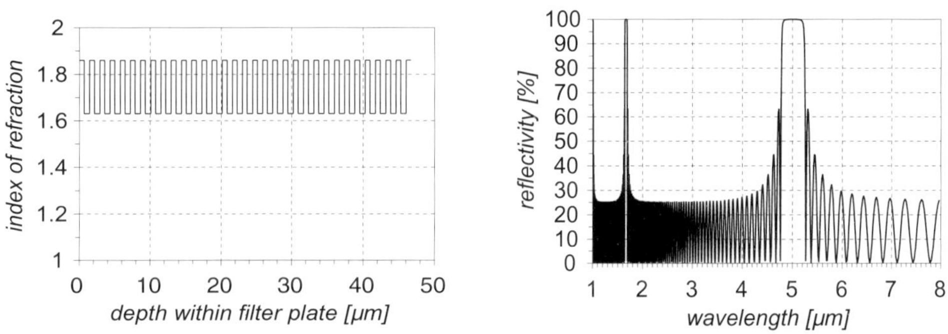

Fig. 6.9 Left: Index profile of a Bragg reflector with 32 double layers. Right: Simulation of the reflectivity of this filter, designed for a center wavelength of 5.0 µm (2000 cm^{-1}). The 3rd harmonic at 1.67 µm can be seen, as well as sidebands and background.

Southwell proposed several improvements to Bragg filters to solve these problems [209-210]:

- The stepwise rectangular index profile is replaced by a sine function. This filter is called rugate filter and suppresses higher order harmonics.
- This sine wave is enveloped with an apodization function to let it run out smoothly at the ends. This suppresses the sidelobes from a finite number of layers.
- The step between the average refractive index of the filter and the refractive index of the outer medium is smoothed with a

polynomial of 5th order. This suppresses the background intensity, similar to an anti-reflective coating.

The polynomial function for the media adaptation layers, also called apodization function, is:

$$n(u) = n_0 + (n_1 - n_0)(6u^5 - 15u^4 + 10u^3) \quad \text{with} \quad u \in [0, 1] \tag{6.19}$$

where u is the normalized layer thickness, n_0 and n_1 the indices of refraction to be linked with a smooth and antireflective transition. The same polynomial is used for the overall envelope function. Fig. 6.10 shows the result of the improved rugate filters. For numerical calculations the sine function has been discretized in 24 sample points.

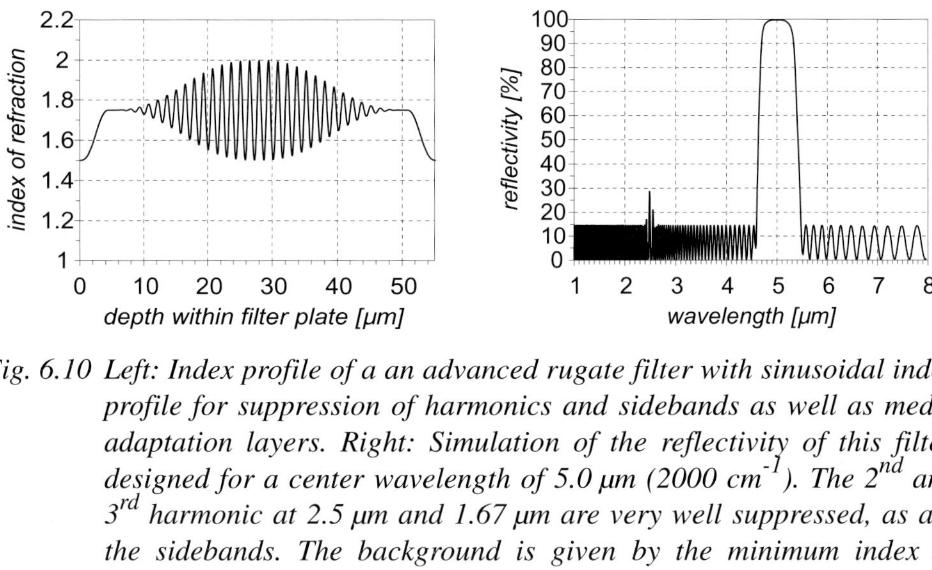

Fig. 6.10 Left: Index profile of a an advanced rugate filter with sinusoidal index profile for suppression of harmonics and sidebands as well as media adaptation layers. Right: Simulation of the reflectivity of this filter, designed for a center wavelength of 5.0 µm (2000 cm^{-1}). The 2nd and 3rd harmonic at 2.5 µm and 1.67 µm are very well suppressed, as are the sidebands. The background is given by the minimum index of refraction that can be fabricated.

6.7.4 Fabry-Pérot bandpass filters

A Fabry-Pérot filter consists of two parallel mirrors, spaced by a $\lambda/2$ layer. It transmits a narrow bandpass where the ideal width is given by the reflectivity R of the mirrors:

$$\frac{\Delta\lambda}{\lambda} = 1 - R \tag{6.20}$$

Two Bragg band-reflectors can be used as mirrors, separated by a $\lambda/2$ layer. For $R = 99.8\%$ this means a spectral resolution of 1 / 500. However, the stopbands that block other wavelengths are not very large, depending on the index modulation amplitude. Fig. 6.11 shows a calculation of a Fabry-Pérot filter. The transmission band is 10 nm wide.

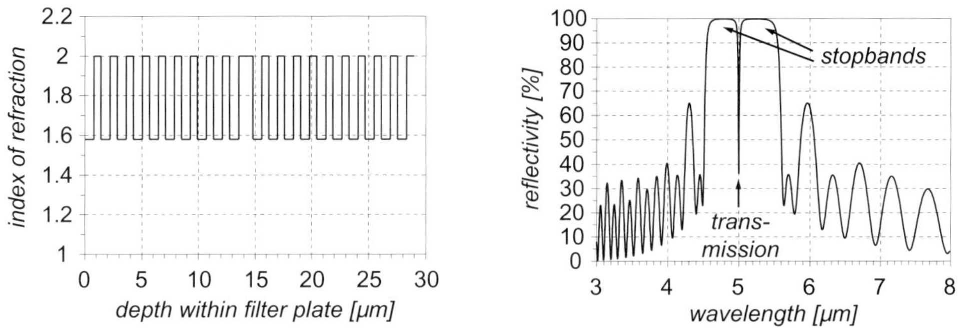

Fig. 6.11 Left: Index profile of a Fabry-Pérot bandpass filter. Right: Simulation of the reflectivity of this filter, designed for a transmission wavelength of 5.0 μm (2000 cm^{-1}). The transmission window is only 10 nm wide. The stopband width is limited.

Advanced rugate index profiles as mirrors do not improve the stopband characteristics compared to simple rectangular profile layer stacks.

6.7.5 Multi band reflectors

A filter spectrum with two or more reflection bands is achieved by additive superposition of two periodic functions with different frequencies. This may associate the analogy of a Fourier transformation between the index profile and the corresponding filter spectrum. Unfortunately, optics is more complicated and a back-transformation to get the required index profile does not exist. The difference is that in a multilayer filter the refractive index is not simply convoluted with the sine and cosine function describing the phase. With growing depth, a part of the lightwave has already been reflected backwards and only a fraction interacts with the lower layers. On the other hand, the lightwave passes some layers several times, see the central λ/2 layer of the Fabry-Pérot filter.

The analogy of a Fourier transformation can help in some cases like the multi band reflector to design a reflective index profile. It is convenient to optimize this profile in Fourier transformed space, with frequencies, amplitudes and phases as open parameters. Fig. 6.12 shows the simulation of a multi-band reflector.

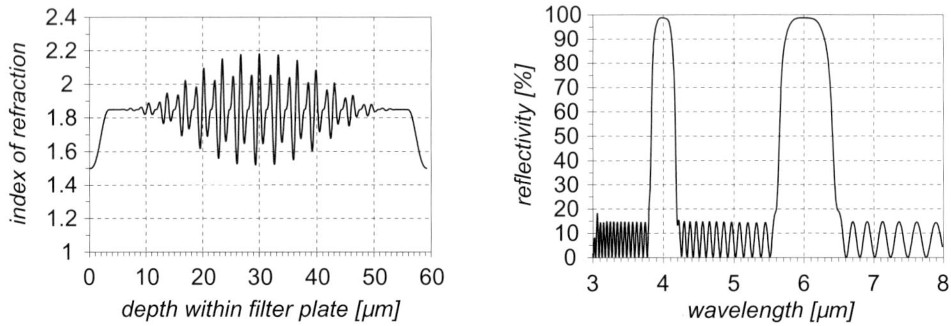

Fig. 6.12 Left: Index profile of a multiband reflector, composed of two modulation frequencies. Right: Simulation of the reflectivity of this filter, designed for reflectivity bands at 4.0 µm and 6.0 µm.

6.7.6 Edge filters

Reflecting a large band can be achieved with a chirped filter, this means a modulation frequency sweep. In the Fourier analogy, a frequency sweep covers a spectral band. When very broad bands are requested, the following effect can be used: If the frequency modulation of a rectangular index profile covers a band such that the end frequency is three times higher than the start frequency, the reflected band overlaps with its own harmonic. The filter will then reflect all wavelengths shorter than a certain value and give an edge spectrum. It can also be used as long wave pass transmission filter. Fig. 6.13 shows the corresponding simulation.

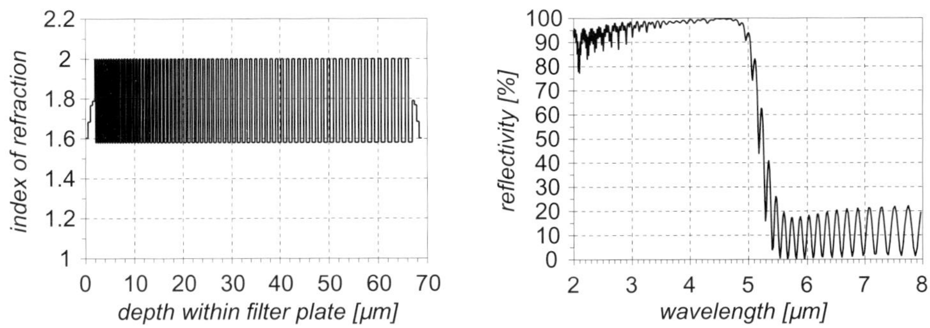

Fig. 6.13 Left: Index profile of long wave pass filter with chirped index profile. Right: Simulation of the reflectivity of this filter, designed for a cut-off wavelength of 5.2 μm.

6.7.7 Angular dependence

The filter spectrum shifts with the light incidence angle θ according to Bragg's law:

$$\lambda_\alpha = \lambda_0 \sqrt{1 - \left(\frac{\sin\theta}{n}\right)^2} \tag{6.21}$$

where λ_0 is the central wavelength at normal incidence that will be shifted to λ_α, and n is the effective index of refraction.

This effect is exploited in the next chapters to make the filter tunable by rotating it mechanically. Although the angular dependence of conventional interference filters is used in some optical devices, this was not reported in literature for porous Si up to now.

The shift is more pronounced for smaller n. As it will be shown in section 6.8.4, the index of refraction can be lowered considerably by total oxidation of the filter to allow large tuning ranges. Oxidation also lowers the reflected background intensity.

6.8 FABRICATION OF OPTICAL FILTER OF POROUS SILICON

6.8.1 Effective media theory

If the pore size is much smaller than the wavelength, porous Si appears homogenous to light and can be modeled with an effective index of refraction n_{eff}. This is sometimes referred as effective medium concept [211]. To create

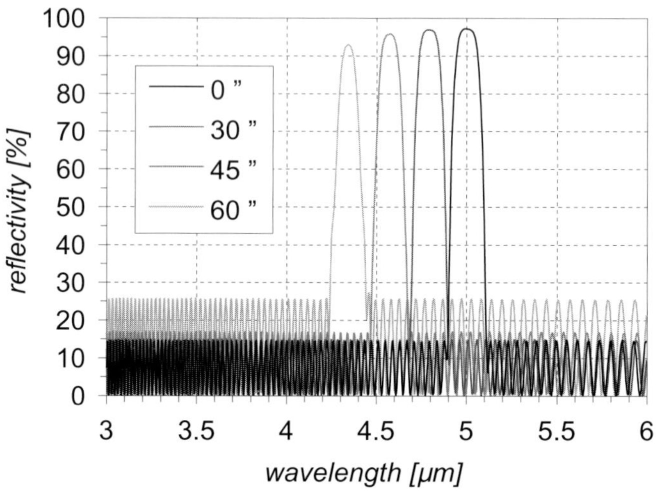

Fig. 6.14 Simulation of the angular dependence of a rugate band reflector. The central wavelength shifts from 5.0 μm to 4.35 μm for incidence angles from 0° to 60°.

optical filters of porous Si with a defined refractive index profile, the relation between refractive index and porosity has to be known. The theory is not trivial because the effective refractive index of porous media is not only a function of porosity (ratio of void to total volume), but it depends also slightly on morphology. Theoretical discussions have been published on that subject [212-216], and there is a Maxwell Garnett, a Bruggemann and a Looyenga formula, using different approaches of effective media approximation. For practical purposes regarding porous Si in this work, the following formula derived by Looyenga as well as by Landau and Lifshitz turned out to give a good compromise between the different theories and fits experimental data well over a large range. It is sometimes referred as "LLL"-formula:

$$n_{eff} = (P + (1 - P)n_{Si}^{2/3})^{3/2} \tag{6.22}$$

P is the porosity and n_{Si} the refractive index of crystalline bulk Si, about 3.85 for visible light and 3.43 for mid infrared, detailed tables can be found in [217]. Put in another form, the formula allows to calculate the necessary porosity for a given n_{eff}.

$$P = \frac{n_{Si}^{2/3} - n_{eff}^{2/3}}{n_{Si}^{2/3} - 1} \tag{6.23}$$

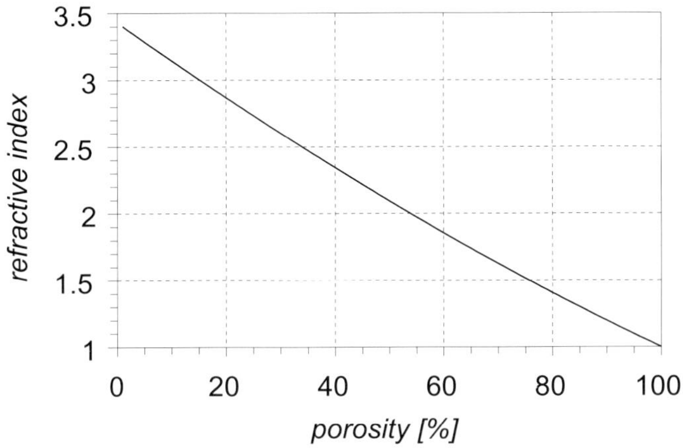

Fig. 6.15 The effective refractive index versus porosity of a porous Si layer in the mid infrared (λ_{bulk} = 3.43).

From Eq. 6.2 one gets the required current density for etching, which finally gives the etch current I as

$$I = j(P(n)) \cdot A \tag{6.24}$$

with the wafer surface area A to be etched. The area increases with time because of underetching, which will be discussed in section 6.8.3.

The electrochemical etch rate is in general

$$\frac{dz}{dt} = \frac{j}{P \cdot v} \cdot \frac{M}{\rho \cdot e \cdot N_A} \tag{6.25}$$

v electrovalence (2.0 to 2.8 for porosification, 4.0 for electropolishing)
M atomic mass per mole, 28.08 g/mol for Si
ρ mass density, 2.33 g/cm^3 for Si
e unity charge, $1.6022 \cdot 10^{-19}$ A·s
N_A Avogadro's constant, $6.022 \cdot 10^{23}$ 1/mol

For Si the second factor on the right side is 1.249 µm cm^2 / As. Typical etch rates are 100 nm / s at a porosity of 55%. If applied to electropolishing, P is set to one, giving typical rates of 600 nm / sec at 2 A / cm^2.

With these equations, the etch current profile for the requested filter can be calculated. Fig. 6.16 illustrates the sequence from optical spectrum to etch current vs. time.

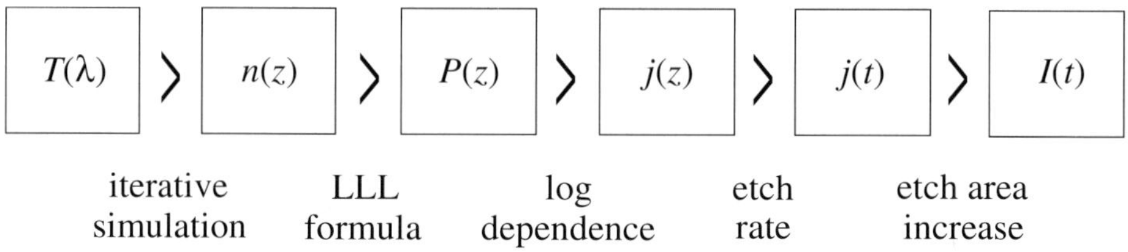

Fig. 6.16 Sequence of calculation steps for modeling porous Si filters.

6.8.2 Lateral homogeneity

Depending on the geometry of the etch cell, the current density at the border of the wafer can be larger than in the center, if e.g. the counter electrode is larger than the wafer. Masking the Si surface with an insulating mask to create microstructures also perturbs the current lines and leads to an increase of current density for small areas and at the border of structures. This problem will be discussed more in detail together with micromachining process technology.

6.8.3 Depth homogeneity

For homogeneity of porous Si layers in depth, two corrections have to be applied, a chemical and a geometrical one.

Correction for HF concentration drop with depth

During the etch process, the HF concentration at the etch front decreases as the exchange of fresh electrolyte through pores is diffusion limited. This leads to higher porosities for the same current density, see section 6.4.4. The problem was reported by Thönissen *et al.* [218] and two solutions have been proposed:

- In situ process control, observing optically the oscillations in reflectivity at a specific wavelength on the surface as the porous layer grows [219-220]. This method is difficult to apply for layers with modulated porosity, which filter the light already.
- Pauses during the etch process allow the electrolyte concentration to regenerate [221]. The problem here is that after every break, the surface charge layer has to build up once more, leading to spikes in the porosity profile.

Both methods were not suitable for the application envisaged. However, it is essential to compensate the depth inhomogeneity, otherwise errors of 20% and more in wavelength result for filters 30 µm thick. The solution developed in this work is to adapt the parameter P_0 with depth in the relation of current density versus porosity (Eq. 6.2). A drop in HF concentration leads mainly to a constant shift of the function *P(j)* to higher porosities. Looking at the diffusion laws, a constant concentration gradient can be assumed, leading to a linear decrease of HF concentration over depth, as experimentally confirmed by Berger [222]. The necessary correction factor was determined experimentally in the following way.

A Bragg reflector typically 50 µm thick was etched to give a reflection band for visible wavelengths. It was detached from the substrate by electropolishing. As the penetration depth of visible light in Si and also in porous Si is limited to a few µm, only the first layers of the Bragg filter contribute to the reflection. If the color on top and on the bottom is not the same, the deviation can be used to determine the necessary correction factor. The central wavelengths λ on top and λ' at the bottom have been measured. In the following it will be shown how the corrected parameter P_0' for the bottom can be calculated. The primed variables refer to the bottom, the unprimed to the top. The wavelength ratio corresponds to that of the optical thicknesses.

$$\frac{n \cdot d}{n' \cdot d'} = \frac{\lambda}{\lambda'} \qquad (6.26)$$

As the electrical etch charge per layer period is constant and assuming that the electrovalence remains constant, the etched volume per layer period is also constant on top and bottom, giving

$$d \cdot P = d' \cdot P' \qquad (6.27)$$

Eq. 6.22 differentiated with respect to *P* gives

$$\frac{\partial n}{\partial P} = \frac{3}{2} \cdot \left(P \cdot (1 - n_{Si}^{2/3}) + n_{Si}^{2/3}\right)^{1/2} \cdot (1 - n_{Si}^{2/3}) \qquad (6.28)$$

which can be used to calculate the relative derivative $\frac{\partial n / n}{\partial P / P} = \frac{\partial \ln n}{\partial \ln P}$.

This allows to calculate the relative change in refractive index from the relative change in porosity by the following equation for small variations.

$$\frac{n'}{n} = \left(\frac{P'}{P}\right)^{\frac{\partial n / n}{\partial P / P}} \qquad (6.29)$$

Together with Eqn. 6.27 the optical thickness at the bottom can be expressed as

$$n' \cdot d' = n \cdot \left(\frac{P'}{P}\right)^{\frac{\partial n/n}{\partial P/P}} \cdot d \cdot \frac{P}{P'} \tag{6.30}$$

Solving for the porosity P' at the bottom gives

$$P' = P \cdot \left(\frac{\lambda}{\lambda'}\right)^{\frac{1}{1-\frac{\partial n/n}{\partial P/P}}} \tag{6.31}$$

For an average porosity of 65%, p^+ Si and $n = 1.86$ for visible light, the exponent is 0.516.

The corrected parameter P_0' at the bottom is calculated as

$$P_0' = P_0 + P\left(\left(\frac{\lambda}{\lambda'}\right)^{\frac{1}{1-\frac{\partial n/n}{\partial P/P}}} - 1\right) \tag{6.32}$$

This calibration can be used to adapt the offset of the function $j(P)$ (Eqn. 6.2) linearly with etch depth. The adapted current density to get the correct porosities will also influence the etch rate, so that the etch times per layer have also to be adapted. The electrical charge $j \cdot t$ per area has to be conserved, assuming again constant electrovalence.

Using this method, the top and bottom filter wavelength have found to match within 1%, compared to 20% deviation for an uncompensated process. The calibration can then be applied also to thick infrared filters. The results show that the drift of process parameters over depth has been canceled successfully. If the optical thickness of a modulation period would vary over depth, the spectral width of a band reflector would broaden similar to the effect of a chirped filter or a deformation by non-uniform strain reported in Ref. [223].

For optimum reproducibility it is necessary to work always at the same temperature (here 20 °C), as this can influence the porosity via the viscosity of the electrolyte. A potential error source are also leakages under the o-rings of the wafer holder that decrease the etch current.

Geometrical correction

The other correction that is necessary for a well calibrated process is of geometrical nature and results from the fact that the interface between porous Si and bulk Si increases with underetching of microstructured surfaces. Assuming isotropic underetching of a given start area A_0 with a perimeter L, at a depth z the area will increase to

$$A(z) = A_0 + z \cdot \frac{\pi L}{2} \tag{6.33}$$

The total etch current must be calculated with this increasing interface area if precise control of porosity is needed. $I(t)$ is then enveloped by a ramp function. Eqn. 6.33 neglects corner points. For dot structures, a quadratic term has to be added.

The open surface area might be difficult to determine or other reasons may make corrections necessary. In this case, the same method of comparing top side and bottom side wavelengths can be applied. Instead of P_0, the slope of area increase will be corrected. The etch time is kept constant. To find the right correction factor for the area and the current, the current density at the bottom j' that appeared in the calibration experiment has been calculated from λ/λ'. The equations change because the electric etch charge per area cannot be assumed as constant any more. The geometric layer thickness at the bottom becomes

$$d' = d \cdot \frac{P}{P'} \cdot \frac{j'}{j} = d \cdot \left(\frac{j'}{j}\right)^{1 - \frac{\partial P/P}{\partial j/j}} \tag{6.34}$$

The influence of a variation of current density to the refractive index can be found by putting Eqn. 6.2 in Eqn. 6.22 and differentiating by j

$$\frac{\partial n}{\partial j} = \frac{\partial n}{\partial P} \cdot \frac{\partial P}{\partial j} = \frac{3}{2} \cdot \left(P \cdot (1 - n_{Si}^{2/3}) + n_{Si}^{2/3}\right)^{1/2} \cdot (1 - n_{Si}^{2/3}) \cdot \frac{\gamma}{j + \zeta} \tag{6.35}$$

which allows to calculate the average refractive index at the bottom as

$$n' = n \cdot \left(\frac{j'}{j}\right)^{\frac{\partial n/n}{\partial j/j}} \tag{6.36}$$

Together with Eqn. 6.34 this gives

$$n' \cdot d' = n \cdot d \cdot \left(\frac{j'}{j}\right)^{1 - \frac{\partial P/P}{\partial j/j} + \frac{\partial n/n}{\partial j/j}} \tag{6.37}$$

and the current density at the bottom j' is

$$j' = j \cdot \left(\frac{\lambda}{\lambda'}\right)^{\frac{1}{1 - \frac{\partial P/P}{\partial j/j} + \frac{\partial n/n}{\partial j/j}}} \tag{6.38}$$

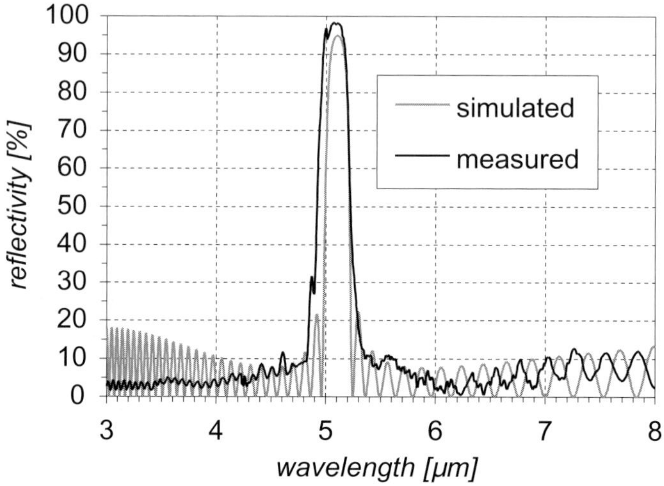

Fig. 6.17 Simulated and measured reflectivity of a rugate bandreflector with a central wavelength of 5.1 µm and a bandwidth of 0.3 µm. The refractive index profile is sinusoidally modulated at constant amplitude and contains additional media adaptation layers.

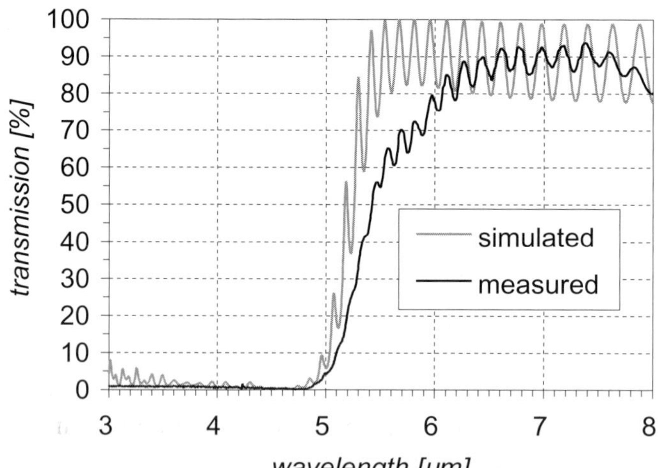

Fig. 6.18 Simulated and measured transmission of a long-wave pass filter that opens at 5.1 µm. The measured filter rejection for $\lambda < 4.8$ µm is below 1%. The refractive index profile is a chirped Bragg reflector like in Fig. 6.13.

For an average porosity of 65% and p^+ Si, corresponding to $j = 162$ mA / cm^2, the exponent is 2.53. The increase factor for the interface area in Eq. 6.33 can now be chosen to match $j' = j$. The correction factor is only valid for the specific wafer mask design for which it was calibrated.

The following Fig. 6.17 and 6.18 show, how theoretical and experimental spectra match together.

6.8.4 Oxidation and aging

HF etches silicon oxides at high rates. The Si surface is hydrogen terminated when the wafer is taken out of the etch bath and rinsed. Silicon oxidizes in air and the first monolayer forms at room temperature after 3 hours. The oxidation mechanism of Si in air at room temperature is entirely different from the thermal oxidation mechanism, the oxide layer thickness grows only logarithmically with time and reaches about 1.1 nm after one year [224].

Porous Si contains a vast internal surface of up to 1000 m^2/cm^3 for p^+ substrates [181] so that oxidation can change the material properties considerably. In optical filters, oxidation leads to a blue-shift, because the index of refraction decreases [225]. To prevent this aging, two solutions have been proposed:

- The porous Si filter is coated with a monolayer of silanes to inhibit oxidation [188]. The method is effective for silanes building a triple bond with Si, like MTCS (methyltrichlorsilane). This compound is chemically much more reactive than HMDS (hexamethyldisilazane), used in micromachining as a primer for photoresists. HMDS passivates the surface only for some days.
- Porous Si can be pre-aged for long term stability by pre-oxidation [226]. It can also be completely converted to porous SiO_2 [227].

The second way was used here, as it is more practical and the only method with long-term data available. Porous Si is entirely oxidized in air at 900°C for 4 hours and converted to porous glass. The porosity modulations of the filter are conserved, but the calibration changes. At temperatures higher than 950°C porous Si densifies because of viscous flow of silica, which is to be avoided because it destructs the multilayer stack. Oxidation makes the filter also transparent for visible light. This opens up more fields of application. Fig. 6.19 shows the reflectivity spectrum of a completely oxidized Bragg filter.

To pre-age the filter, it needs not to be oxidized completely because the oxide thickness will only grow logarithmically with time. A process step at high temperature may not be compatible with other materials used in a microsystem.

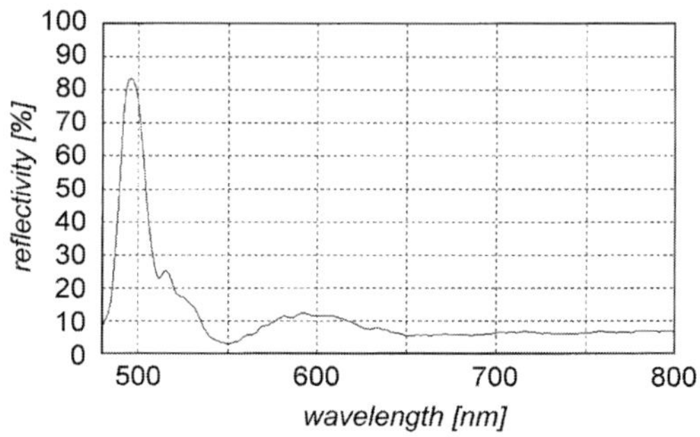

Fig. 6.19 Reflectivity spectrum of a Bragg reflector of 183 periods of density modulated oxidized porous Si at normal incidence.

Instead, a treatment in oxygen plasma oven for 10 min showed sufficient pre-oxidation. In this case only a small fraction of porous Si is oxidized and the shift in optical properties can be corrected by assuming a smaller refractive index for n_{Si} in Eq. 6.22.

For complete oxidation, the oxygen atoms need additional space in the porous Si structure and the porosity decreases. The molar volume of SiO_2 is 2.26 times higher than that of Si, see Table 6.1. If the initial porosity is below the critical value of 56%, the pores at the surface will clog and prevent complete oxidation. The porosity after oxidation P_{ox} can be calculated as

$$P_{ox} = 1 - 2.26 \cdot (1 - P) \tag{6.39}$$

Fig. 6.20 illustrates the porosity range suitable for oxidation. The restriction of the initial porosities to a range of 56 to 75% is not severe. Porosities lower than 56% are not very convenient in practice anyway, they lead to long etching times (compare Fig. 6.6) and larger depth inhomogeneities because of reduced electrolyte exchange and additional electroless chemical dissolution of Si in HF. The final porosity P_{ox} between 0% and 43% has to be used in Eq. 6.22 together with the bulk index of refraction of glass to calculate the index of refraction of entirely oxidized porous Si.

Table 6.1 Material properties of Si and SiO_2.

Material	ρ [g/cm^3]	M [g/mol]	$V_M = M/\rho$ [cm^3/mol]	n_{bulk} at 600 nm
Si [100]	2.330	28.08	12.06	3.95
SiO_2	2.2	60.1	27.3	1.46 (glass)

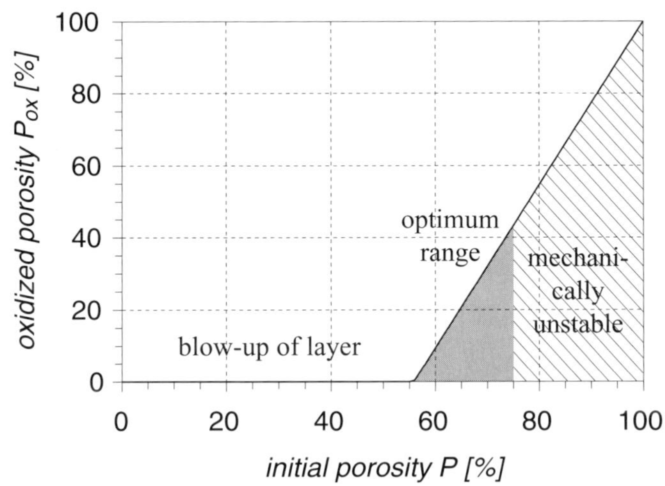

Fig. 6.20 Porosity before and after oxidation. Layers with less then 56% initial porosity do not provide enough free space for integrating all oxygen atoms.

6.9 SUMMARY

Density modulations in porous Si enable the fabrication of optical filters on Si wafers by fast and inexpensive electrochemical etching. The technology is compatible with Si micromachining and allows to integrate optical filters in MEMS. Compared to conventional layer deposition methods porous Si offers more degrees of freedom for optical design because of the possibility to fabricate arbitrary gradients and profiles of the index of refraction.

Different filter types like Bragg-mirrors, Fabry-Pérot bandpasses and edge filters have been designed, simulated and realized. Rugate filters and anti-reflective media adaptation layers improve the filter characteristics by suppressing the background and sidelobes of the filter spectra. Corrections of etch parameters were calculated and applied to compensate the decrease of HF concentration during etching. This calibration improves considerably the homogeneity of the porosification process over the layer depth and leads to better optical characteristics compared to porous Si filters reported previously.

The wavelength resolution of the filter is 20 nm at a central wavelength of 600 nm, which corresponds to the color resolution of the human eye. Infrared Bragg mirrors with sharp reflection bands have been presented, as well as edge filters with a background suppression reduced to less than 1%.

7 MICROMACHINING USING POROUS SILICON

Scheele and many later investigators, including Davy, Gay-Lussac, Lavoisier, and Thenard experimented with hydrofluoric acid, some experiments ending in tragedy.
[128]

7.1 GOALS FOR THE TECHNOLOGY

Porous silicon was proposed since 1994 for micromachining, but at first only for sacrificial layers [193-198]. Lehmann considered it as new base material for MEMS using the high internal surface of macropores in 1996 [228]. Tjerkstra combined porosification and electropolishing of Si to form free-standing layers of porous Si to separate concentric microchannels [229] and uses isotropic electropolishing to micromachine buried microchannels [230]. All the devices have in common that they are mechanically static.

Ohji *et al.* etched electrochemically micrometer sized trenches in Si and connected them at the bottom by changing the etch parameters via illumination or current density [231-233]. The released trenched structure is proposed as accelerometer [234]. The device can flex by some microns by external stimulation, but it contains no actuator.

The goal for the new technology presented here was to integrate a porous Si multilayer with a microactuator and to bring the optical filter plate in motion for a new and original MOEMS device. A new micromachining technology has been established, that is

- simple in the sense that it needs as few process steps as possible,
- compatible with standard cleanroom processes,
- flexible to allow fast design improvements and testing.

In this chapter, the toolbox of materials and technologies for porous Si surface micromachining will be described, as well as the design rules for MOEMS using this technology [235-240].

7.2 Metal masks

Metals that resist HF can be divided in two groups:

- gold (Au) and platinum (Pt) are chemically inert to withstand HF,
- copper (Cu), nickel (Ni), lead (Pb) and other materials protect themselves with a dense oxide layer.

Cu preforms poorly as mask during electrochemical etching of Si. Direct electrical contact with the wafer set the Cu layer to an anodic potential that accelerates its oxidation drastically compared to purely chemical etching.

An evaporated Au layer with an adhesion layer of Cr gives better results. It is structured by photolithography and wet etching in a potassium iodine (KI, I_2) solution. Au withstands standard HF electrolyte well for 30 min. The major problem is that the Cr layer slowly dissolves and detaches the Au layer. Au cannot be deposited directly on Si, it would peal off. The explanation for the degradation of the Cr layer is again found in electrochemistry. Even if no external voltage is applied (or the Cr / Au layer is insulated electrically from the Si substrate), the two metals form a local element with a difference in standard electrode potentials of 2.242 volts. This accelerates the oxidation of Cr as anode. The effect can be reduced with a sandwich structure of Cr / Au / Cr, where Au is only in contact with the electrolyte at the border of the structured layer [249].

Pt is difficult to microstructure for layers as thick as 1 µm. It cannot be etched using photoresist masks because the polymer dissolves faster than Pt in Pt-etch baths. The usual lift-off technology using photoresists with overhanging borders does only work for a maximum thickness of about 500 nm. Further, the same adhesion layer problem as with Au exists.

A good solution for a metal mask problem is a combination of a Ni and a Cr layer that differ by only 0.487 volts in the series of standard electrode potentials. Both metals are evaporated and structured by photolithography and wet etching. Ni is etched in a heated solution of ammonium persulfate ($(NH_4)_2S_2O_8$) and iron(III)chloride ($FeCl_3$). To achieve good etching uniformity and low underetch of the photoresist mask, the bath can be equipped with a nitrogen (N_2) bubble generator. For a 1 µm Ni layer the underetch of the photoresist mask is reduced to 3 µm compared to 10 µm without N_2 bubbles.

Alternatively, Ni can be electroplated. A seed layer of Cr / Au provides the electric contact. Standard positive photoresist masks this layer where no Ni is wanted and leaves structures open for galvanic deposition of Ni from a $Ni(SO_3NH_2)_2$ bath. 1 µm Ni is deposited within minutes. The photoresist is removed, Au and Cr is etched away where no Ni is deposited. Resolutions down to 1 µm can be achieved. Ni resists HF considerably better than Cu and Cr and can be recommended as HF compatible material for micromachining of porous Si.

7.3 NITRIDE MASKS

Silicon nitride (Si_3N_4) works well as mask on Si to define porous Si structures, even down to sub-micrometer resolution [250, 251]. This material was chosen as electrically insulating mask layer and showed good and reliable results. It is structured by standard photolithography and dry reactive ion etching (RIE) with inductively coupled plasma (ICP) in sulfur hexafluoride (SF_6) or hexafluoroethane (C_2F_6). An optical etch stop detection avoids etching into the Si substrate.

The best results are obtained with low stress LPCVD (Low Pressure Chemical Vapor Deposition) Si_3N_4, and layer thicknesses between 120 nm and 500 nm. Si rich nitride, also referred as $Si_{3+x}N_{4-x}$, adapts better to the crystal lattice constant of the Si substrate. Stoichiometric Si_3N_4 rolls up at the corners when it is underetched because of layer stress.

7.4 FREE-STANDING POROUS SILICON FILMS

Layers of porous Si can be detached from the substrate wafer by sub-surface electropolishing, see section 6.5. Therefore, the current density is increased above the critical current density directly after porosification. This method was originally used to characterize the bottom side of porous layers, to look under them [191]. Fig. 7.1 illustrates the current density versus time. The porosification current can be modulated to create porosity profiles for multilayer optical filters.

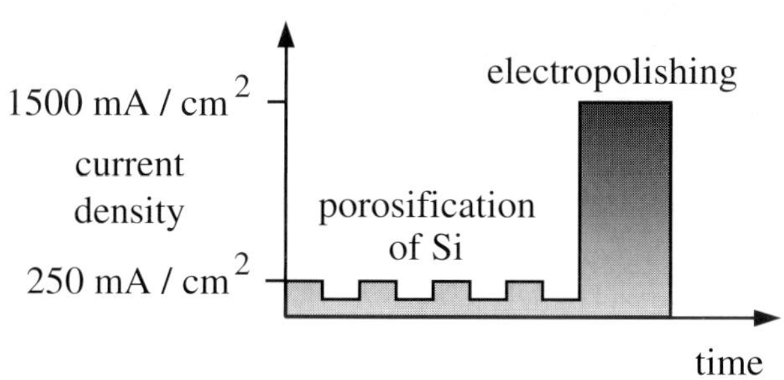

Fig. 7.1 Schematic diagram of current densities during the phases of porosification and electropolishing of Si.

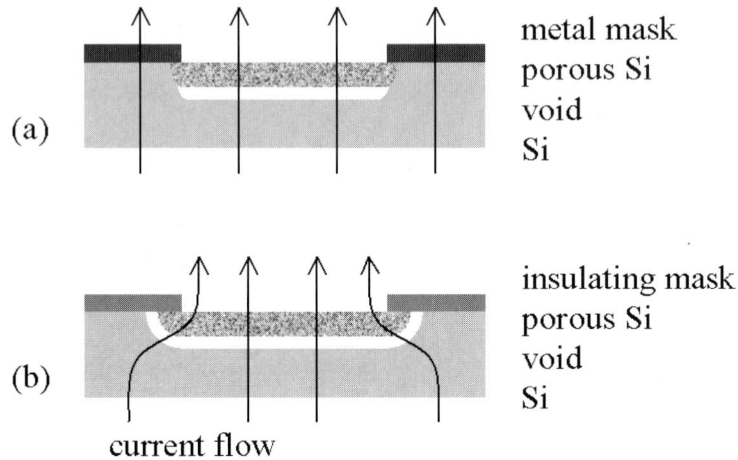

Fig. 7.2 Formation of porous Si with subsequent electropolishing gap. (a) structured using a metal mask, (b) structured using an insulating mask. Lines of electrical current flow are indicated.

Depending on the masking technique used, the porous layer can either be held at the border or released. Metal masks result in an almost homogenous current flow during the electrochemical etch process. The current can leave through the mask and will flow mainly perpendicular to the wafer surface, Fig. 7.2 a. The layer thickness of porous Si will show good uniformity [253]. The lateral underetching is less than the layer thickness, this means anisotropic etching. Electropolishing at moderate current densities will release the layer from the substrate but not from the borders. If the generation of gas bubbles during sub-surface electropolishing is too violent the porous layer will break. A porous layer suspended at the borders can serve as a membrane. Thicknesses ranging from less than 1 μm to more than 50 μm are possible.

In the case of insulating masks, the current coming from the backside plane is forced to squeeze through the mask openings, leading to a lateral current flow, Fig. 7.2 b. The current density is increased at the sides of the mask openings and the lateral underetching is nearly isotropic. As a consequence of lateral current the subsequent electropolishing will also release the sidewalls of the porous layer. The porous part is then only held by the mask in the underetched zone and has no contact to the substrate anymore. It was found in agreement with Ref. [253] that the porous layer is thicker at the borders of the mask openings and also that electropolishing is deeper there, resulting in a W-shaped profile. This effect is less pronounced for highly doped, low resistive p^+ Si because the current density can equalize better over the wafer.

On top of a Si_3N_4 mask a metal layer can be deposited and structured that serves as resistive heater wire for a thermal bimorph actuator or as electrodes for actuators or sensors. Conductive tracks of Cr / Ni can be underetched with porous Si structured by a Si_3N_4 mask. If Ni is directly evaporated on Si, no adhesion layer is necessary, because it reacts and forms NiSi [254]. This is not the case for a Si_3N_4 surface, where Cr is needed to promote the adhesion.

7.5 Mask removal

The problem of a chemically resistive mask is that it is also hard to remove. Dry etching of Si_3N_4 after the electrochemical etching destroys the porous Si plate. Concentrated phosphoric acid (H_3PO_4) at 160 °C etches Si_3N_4 selectively with respect to Si, but the installation needs a condensate reflux bath and it is complex to ensure safety.

Low stress Si_3N_4 is also etched in HF at a rate of 6 nm / min. For bulk Si the HF etch rate of less than 1 nm / min can be neglected. However, in porous Si the wall thickness of the internal nano-structure would change within the 30 min needed to dissolve the Si_3N_4 mask and the optical properties would degrade. The internal structure of porous Si would become thinner by chemical etching.

The solution developed in this work is to leave the wafer in HF electrolyte after etching for 30 min and reverse the polarity of the etch current, forming porous Si on the backside of the wafer. The reverse potential puts the porous Si on the frontside at a cathodic potential that prevents it from oxidation. It works like a cathodic anti-rust protection and conserves the optical filter plate during 30 min in HF without any change of the optical properties. The method has been patented [240].

7.6 Thermal actuator design

We present a new process for the fabrication of a layer sandwich for a thermal bimorph or trimorph actuator by surface micromachining using porous Si and a metal heater. An intermediate Si_3N_4 layer insulates the metal from the substrate; at the same time it defines the regions to porosify.

It would be possible, but it is not practical to porosify first and deposit the metal layer afterwards. At the end the filter has to be released from the substrate, preferably by sub-surface electropolishing under the porous layer. The wafer would have to be processed in the HF bath a second time. It is more practical to porosify and electropolish in one process step at the end. Therefore, the following process sequence has been developed.

Si_3N_4 is deposited on the Si wafer, then the metal layer (Cr/Ni or Cr/Au/Ni) on top. It is structured by photolithography and etching to form heater wires and the upper layer of the multimorph actuator. Then the Si_3N_4 mask is structured to define the areas to porosify. It can obviously only be opened where there is no metal. The optical filter and actuator are electrochemically porosified and released by electropolishing in one last process step.

To release the actuator arms from the substrate the metal tracks have to be underetched with porous Si. However, the mechanical requirements for the porous Si filter plate and for the actuator are different, see Table 7.1.

Table 7.1 Structural requirements for the filter plate and the actuator.

Filter	**Actuator**
flat rigid plate	pre-bent flexible cantilever
thick (≥ 30 μm)	thin (typ. 3 μm)
released from substrate	suspended to substrate at one end

Mask perforation

These different structural properties are generated with only one photolithography mask level. Flexible cantilevers are achieved by perforating the Si_3N_4 mask with square holes of typically 20 μm sidelength and 50 μm period (Fig. 7.3).

Isotropic electrochemical etching creates an array of bumps of porous Si under the perforated region. The bumps are linked when the lateral underetching of two adjacent perforation holes meets and the structure becomes a cantilever (Fig. 7.4). The average current density at the porous Si / bulk Si interface under a perforated region after underetching is lower than in regions where the mask is completely open. This makes the suspension thinner and more flexible than the filter plate.

Directly after porosification the current density is increased to $2\ A / cm^2$ to electropolish the interface between porous Si and bulk Si and release the arm from the substrate. It is anchored by the Ni layer where the heater wires join the bondpads. The shape has been chosen to run out smoothly to prevent excessive

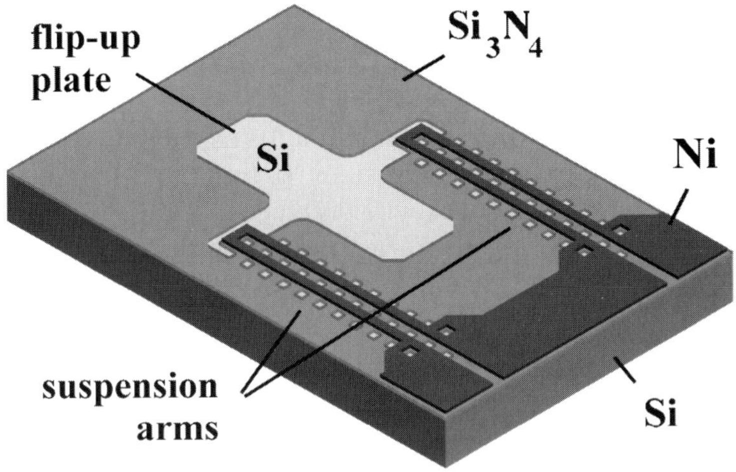

Fig. 7.3 *The Si wafer with two mask levels, prepared to be electro-chemically etched. The Si_3N_4 mask defines the area of the flip-up plate. The Ni tracks on top serve as heater wires for two thermal bimorph actuator arms. Chip size: 3 x 4 mm^2 typically.*

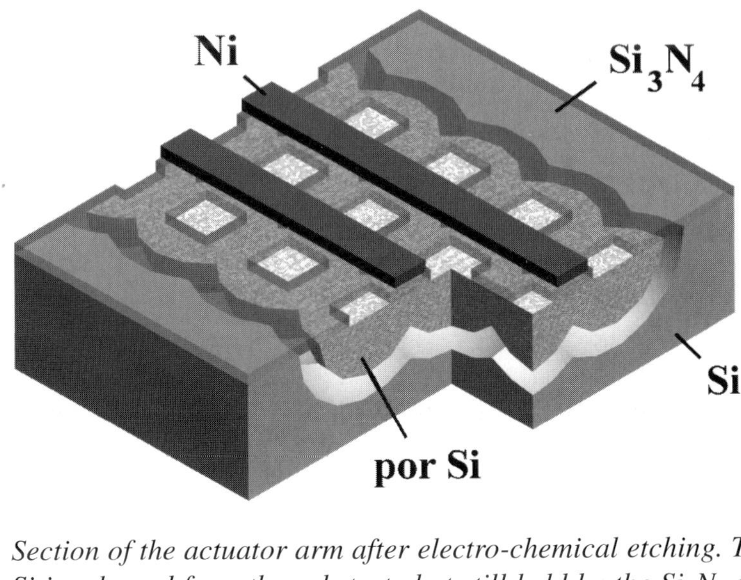

Fig. 7.4 *Section of the actuator arm after electro-chemical etching. The porous Si is released from the substrate but still held by the Si_3N_4 mask, here shown transparent. It will be dissolved to release the structure. Typical dimensions of openings in Si_3N_4: 20 x 20 μm^2, Ni tracks: 25 μm width.*

strain in the corners. The decreased current density at the end of the beam also inhibits electropolishing at the very end, so that the arm is linked by a bridge of porous Si to the substrate. The other end of the actuator arm holds the filter plate.

The optimum dimensions for spacing and size of the perforation holes have been determined experimentally. The width of the metal tracks on top is limited by the hole spacing, defined as repetition period of the array minus hole sidelength. A minimum metal track width is required for stability, otherwise the arms become too fragile. If the tracks are too large underetching and linking the bumps becomes difficult, especially as the overlap of the bumps shall be well defined. An important parameter is the average mask opening in the perforated region, given by:

$$opening = \left(\frac{sidelength}{sidelength + spacing}\right)^2 \tag{7.1}$$

It was found that the optimum opening is between 16% and 25%. Smaller values give not enough current density to underetch the perforation array. Larger values lead to almost the same arm thickness as the filter plate and make the cantilever too rigid. Fig. 7.5 shows the optimum design window.

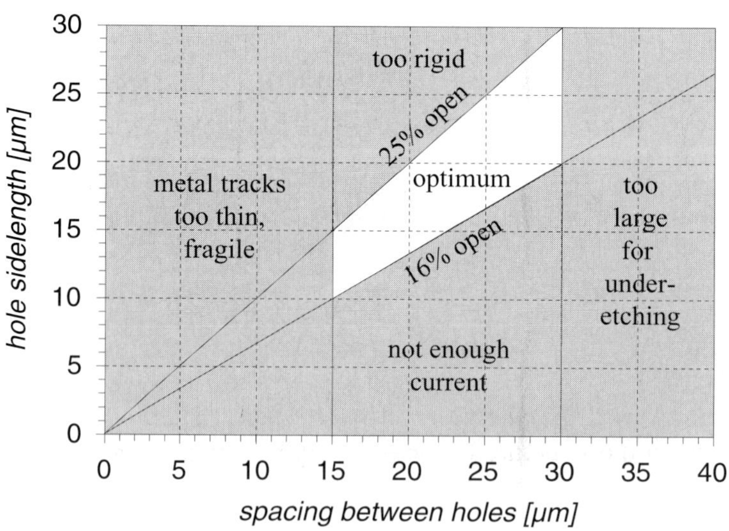

Fig. 7.5 Optimum spacing and sidelength of perforation holes in the Si_3N_4 mask. The optimum design window for the actuator was determined experimentally.

Fig. 7.6 shows a detail of an actuator arm seen from the bottom. The porous Si bumps from underetching of the perforation holes intersect at the borders. When the porosity gradient is too high, the structure can get fissures like it can be seen in the shell of the bumps. A close-up of two intersecting bumps is shown in Fig. 7.7.

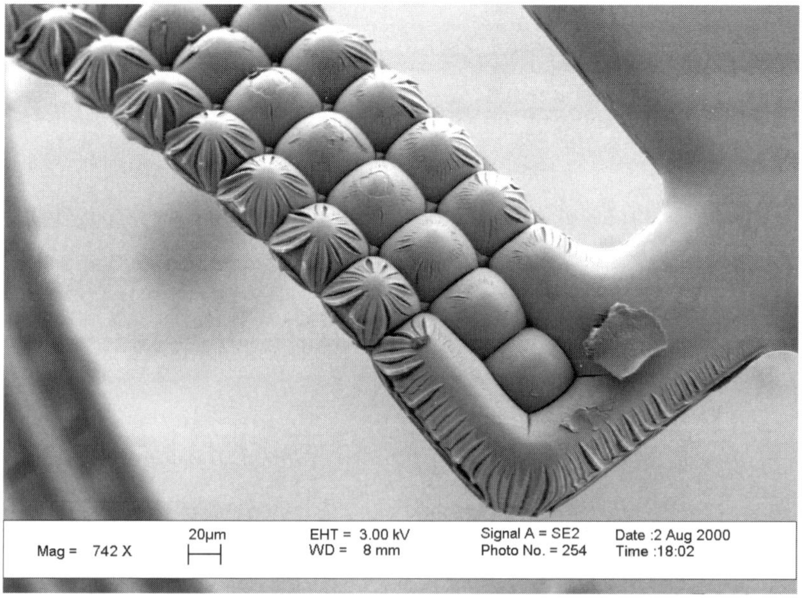

Fig. 7.6 The actuator arm seen from the bottom. The bumps of porous Si became brittle because of a high porosity gradient.

Alternative actuator geometries

Instead of a cantilever shape the actuator can also have the form of a ribbon that suspends the filter plate at its bottom edge. Instead of a rectangular perforation array a hexagonal array was also investigated, Fig. 7.8 shows the result. As expected the structure became too rigid to flex so that this option has been dropped.

7.6.1 Calculation of optimum layer thicknesses

Geometry

Using the described process, the actuator layer sandwich makes a trimorph cantilever of Ni on Si_3N_4 on porous Si. The geometry is complicated to model

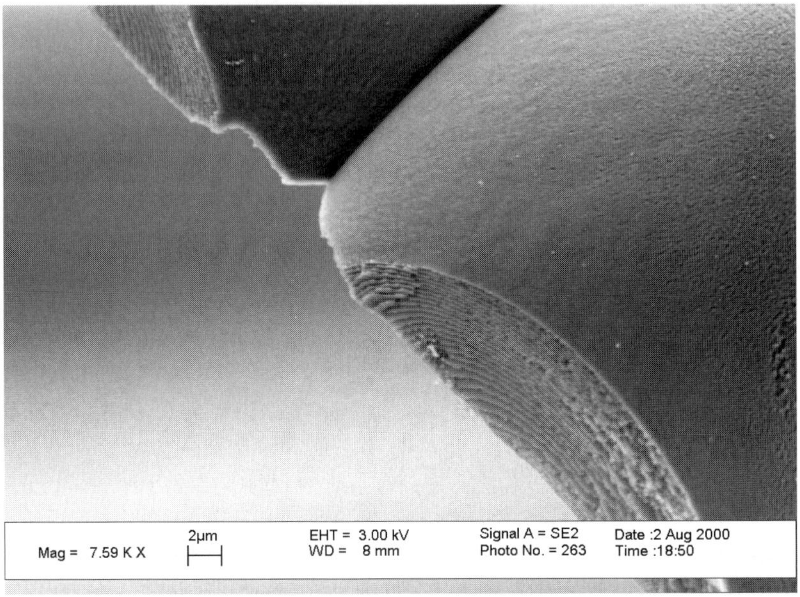

Fig. 7.7 *Close-up of two bumps of an actuator arm. At the cleaved edge the concentric shell structure of porosity modulation becomes visible.*

Fig. 7.8 *SEM picture of a plate of porous Si with a hexagonally perforated mask region (bumps), seen from the bottom. The lower edge has been cleaved to look at the bottom side of the plate. The negative profile of the bumps can be seen on the substrate on the bottom of the picture.*

because the Ni and Si_3N_4 tracks are not as wide as the porous Si under it, and the thickness and width of the porous Si layer is inhomogeneous. The Young's modulus of the porous Si layer is not homogeneous either because the porosity varies. It is modulated for the optical multilayer filter and there is a superimposed gradient because of the following reason: The interface porous Si / bulk Si increases faster under the perforated area than under the optical filter area for which this effect is corrected (see section 6.8.2).

A finite element model (FEM) simulation of the bump structure would be feasible in principle, but the numerical effort is enormous because the actuator contains in the order of a hundred bumps that consist each of the order of a hundred to thousand concentric half-shells of porous Si with different material properties, each of them would have to be meshed. The results would not even be very reliably as the Young's modulus of thin films of porous Si is not known well. Moreover the material properties of thin evaporated or electroplated films are difficult to determine and they vary with temperature [146].

Instead an analytical model based on an effective thickness and width of the arm, averaging the geometry, has been used to reduce it to the analytic solutions developed in chapter 2 (Fig. 7.9).

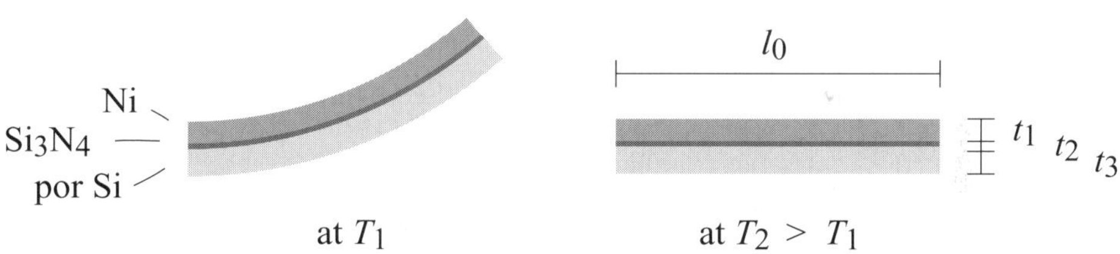

Fig. 7.9 Principle of a thermal trimorph actuator. Ni has a higher thermal expansion coefficient than Si_3N_4 and porous Si. When the beam is heated, it bends down. The initial curvature is due to residual stress from Ni layer deposition and partial oxidation of porous Si.

This model of a trimorph beam can explain already most of the characteristics of the actuator and gives hints what fabrication parameter will influence the device characteristics in what way. The material properties are given in Table 7.2, they are taken from different publications [138 (chap. 4.1 and 4.3), 128, 139-147]. Some values vary strongly with the group reporting. When the thin film property of a material was not available, the bulk property has been used.

Due to the fabrication process the porous Si layer of the actuator arm is undulated in thickness (see Fig. 7.4) so the question arises if the structure can be

Table 7.2 Material properties (at 300 K) for the calculation of the thermal bi- and trimorph actuator. References are given for special measurements.

Material	Young's modulus E[GPa]	thermal expans. coeff. $\alpha[10^{-6}/K]$	mass density ρ $[g/cm^3]$	specific heat capacity c_{th} [J/kg·K]	thermal conductiv. λ_{th} [W/m·K]	electrical resistivity ρ_{el} [μΩ·cm]
Ni (bulk)	208	12.7	8.91	443	90.5	8.7 [145], 6.84, 10[128]
Ni (e-plated)	≈135 [146]	8 .. 16 [146]				
Au	78	14.2	19.3	129	317	2.2
Cr (bulk)	279	10.0	7.15	449	93.7	12.5
Cr (evap. film)	280 ± 10	5.0				≈ 130 [256]
Si_3N_4	290	2.9	3.1	710	19	≈ ∞
low stress Si_3N_4	95 ± 10 [143]		2.86 [143]			≈ ∞
Si [100]	130	2.3	2.33	713	156	$p^+ \approx$ 15k
por Si (P=50%)	≈ 20 [138 (4.1)]	≈ 2.3	1.16		2 .. 80 [140, 141]	> 1M
por SiO_2 (P=20%)	< 75	≈ 0.3	≈ 2.1		2.7	≈ ∞

described by an effective thickness and to which averaging the value for porous Si refers. Between two perforation holes on the actuator arm the porous Si intersects only by some microns, determined by the underetching time during porosification. In the center of a bump under a hole the thickness is about half the perforation spacing period, typically 25 μm. The same thickness modulation is found along the cross section of the cantilever.

One approach is to regard only the thin parts as flexible in the regions where the bumps intersect and the center part of the bump as stiff. The calculated value for porous Si corresponds then to the minimum thickness. Only the flexible sections contribute to the effective actuator arm length. The real cantilever length has to be longer because it also contains stiff sections.

As mentioned, the Young's modulus of the porous Si layer is not homogeneous. The current density drops over time in the actuator region as the underetched surface increases. This means that the porosity decreases and the shell of the bump is more solid than the core under the perforation hole, where the current is concentrated (Fig. 7.10). The Young's modulus of porous Si increases with decreasing porosity so that it can be assumed that the outer parts contribute most to the layer stiffness. Therefore the model could be reduced to the shell of the bumps. This leads to the conclusions that the influence of thickness modulation is less severe than it looks at a first sight.

In practice it was observed from the actuation angles obtained (see chapter 4) that the value for the effective thickness for the porous Si layer used in the analytical calculations corresponds rather to the minimum thickness at the intersections of the bumps.

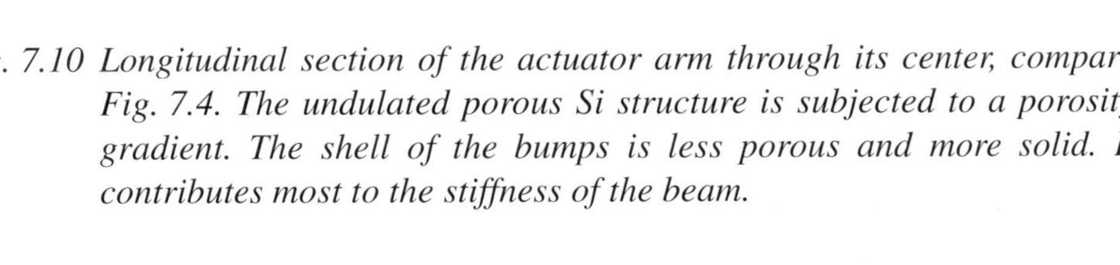

Fig. 7.10 Longitudinal section of the actuator arm through its center, compare Fig. 7.4. The undulated porous Si structure is subjected to a porosity gradient. The shell of the bumps is less porous and more solid. It contributes most to the stiffness of the beam.

To find the optimum combination of layer thicknesses for maximum deflection at given temperature, Eq. 2.18 was numerically plotted with t_1 and t_2 as open parameters and t_3 fixed. An average temperature increase of 400 K over the ambient temperature was assumed, which was found to be the maximum value for long term operation. Fig. 7.11 shows the bending curvature of the trimorph with the porous Si layer kept constant at 3.0 μm effective thickness. As expected the thickness of the Si_3N_4 layer has not much influence. It was fixed to

200 nm – thick enough to withstand the HF electrolyte for porosification and not too thick to dissolve it in a reasonable time of about 30 min afterwards.

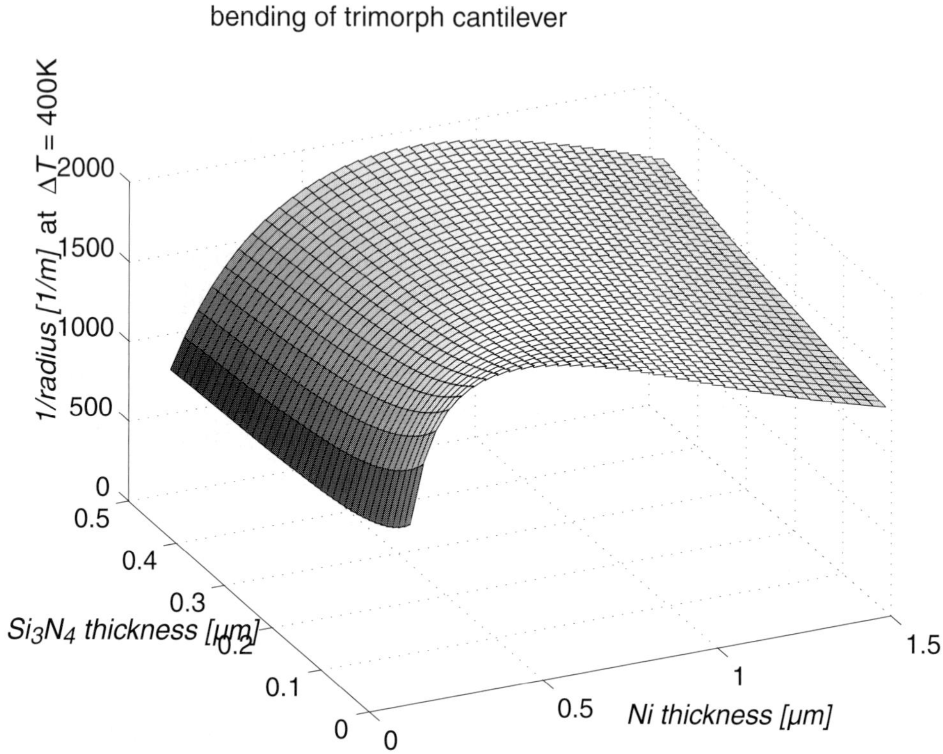

Fig. 7.11 Bending curvature (reciprocal radius) of a trimorph cantilever of 3.0 μm porous Si and Si_3N_4 and Ni of variable thickness. The thickness of Si_3N_4 has minor influence, the optimum thickness for Ni is 0.8 μm.

In the next plot the Si_3N_4 thickness was set to 200 nm and the reciprocal curvature radius was plotted versus the thicknesses of the porous Si and Ni layer, see Fig. 7.12. Thinner layer combinations lead generally to higher curvatures. For a porous Si thickness of 3.0 μm the bending radius does not vary much around the optimum Ni thickness of 0.8 μm. Experimentally is was confirmed that a Ni thickness of 0.8 to 1.0 μm is a good value for high actuation angles and robust microstructures. Thin layer combinations give higher actuation angles at the same heating power. The choice of thicknesses is a compromise between actuator efficiency and robustness. When the Ni thickness is less than 0.5 μm the structures become very fragile.

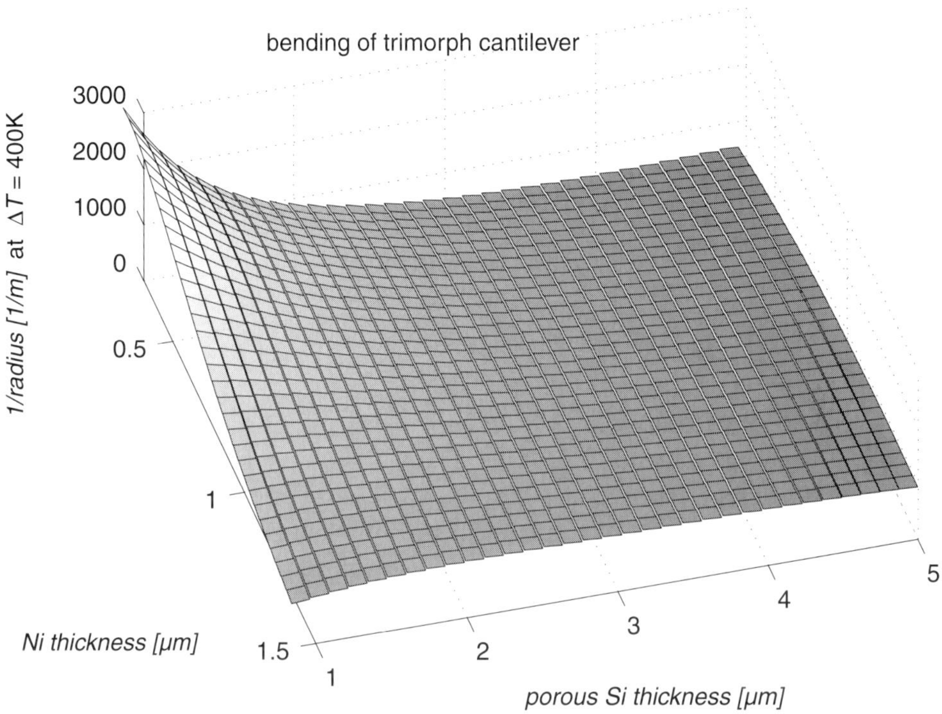

Fig. 7.12 Bending curvature (reciprocal radius) of a trimorph cantilever of porous Si of a thickness between 1.0 and 5.0 µm, 200 nm Si_3N_4 and a Ni thickness ranging from 0.3 to 1.5 µm. Thinner combinations lead to higher curvature but they are more fragile mechanically.

Temperature limits

The mechanical deformation is in first order directly proportional to the temperature change. At high temperatures the thermal expansion coefficient can increase, e.g. for Ni from $12.7 \cdot 10^{-6}$/K at 300 K to $16.8 \cdot 10^{-6}$/K at 800 K [128]. This will increase the actuation angle.

Young's modulus goes down with temperature. If this applies to all materials with the same factor, the net effect is zero because only ratios of Young's moduli appear in Eq. 2.15 and 2.18.

The maximum possible beam temperature is given by the material degradation. Upper limits are surely the melting points of Ni (1453 °C), Si (1410 °C) and Si_3N_4 (1900 °C). The melting point of porous Si is slightly lower than that of the bulk material because of surface melting effects. Experiments in an oven showed that Ni forms an oxide in air at 800°C, so that this should be the peak limit for operation. Envisaging the device application as infrared spectrometer, higher temperatures would also lead to considerable infrared radiation and add a signal background. The empiric value for the maximum average temperature of the actuator for long-term operation is 400 °C.

7.6.2 Calculation of beam length

The resulting actuation angle φ is the integral of the curvature along the beam:

$$\varphi = \int_0^l \frac{1}{r(x)} dx = \int_0^l \frac{\Delta T(x)}{\xi} dx = \frac{l \cdot \langle \Delta T \rangle}{\xi} = l \cdot \langle \frac{1}{r} \rangle \tag{7.2}$$

where the right side of Eq. 2.18 except ΔT is expressed in the constant ξ. $\langle \Delta T \rangle$ is the average temperature in the beam. This means that the temperature distribution does not change the actuation angle φ at the tip. Only the form of the cantilever changes with the temperature distribution. If the tip is hotter than the anchor point the tip position will move less up and down than at uniform temperature distribution.

The optimum set of layer thicknesses and beam length for a given actuation angle is not affected by the temperature distribution. For a trimorph layer stack of 3.0 μm porous Si, 200 nm Si_3N_4 and 0.8 μm Ni the bending curvature is 625 μm at $\Delta T = 400$K. This means with Eq. 7.2 that for 90° actuation angle a beam length of 980 μm is required. Regarding the undulated geometry of the porous Si layer as discussed above, this value refers to the sum of flexible sections of the beam, so the total beam length should be greater. This effect lowers the actuator efficiency because rigid sections have to be heated, too. It was found that for 90° actuation a length of 1600 μm is needed.

7.6.3 Electrical heater resistance

For the calculation of the electrical resistance R of the heater, the Ni layer and in case of electroplating also the 50 nm Au seed-layer have to be considered. The 20 nm thick Cr layer can be neglected because of its higher specific resistivity. In general

$$R = l_{tot} \cdot \left(\sum_i \frac{A_i}{\rho_i} \right)^{-1} \tag{7.3}$$

where l_{tot} is the wire length, A_i the cross section area of layer i and ρ_i its specific resistance. The wire has twice the arm length for every one of the two arms. Assuming an arm length l of 1.6 mm (see design in the following section) this results in l_{tot} = 6.4 mm. The wires are typically 25 μm wide. With the values of Table 7.2, a Ni thickness of 0.8 μm and a Au thickness of 50 nm, the total resistance is 22 Ω.

7.7 MECHANICAL FILTER PLATE SUSPENSION

7.7.1 Shape and size of filter plate

Most MOEMS devices use rectangular mirrors or filter plates. However, the optical beam has mostly a round cross section, so the optical element can have as well rounded corners or a disk or elliptical shape. This saves mass of the plate, increasing the speed of the device. When the beam hits the filter not at normal incidence, the projected area is stretched. For our application with incidence angles of up to 60° the rounded shape has to have an aspect ratio of 1 : 2.

The required size of the filter plate depends on the application. For fibre optics as widely used for telecommunication, the core diameter is typically 125 μm for multimode and 5 μm or less for monomode fibres. Photodiodes with sensitive areas of some 100 μm sidelength are available. Therefore, the smallest filter areas designed were 270 by 700 μm^2.

For infrared gas sensing, the detector areas are usually larger. The smallest thermopiles are about 70 by 750 μm^2, pyroelements often come as 2 x 2 mm^2 and more. The largest filters designed were 2.6 x 3.0 mm^2. The filters become more fragile the larger they are. Reasonable sizes were found to be 1.0 x 2.0 mm^2.

7.7.2 Ways of suspension

In a first design, the filter plate was suspended via an actuator ribbon to the substrate, see Fig. 7.13 A. Although this geometry is mechanically robust, it has several drawbacks:

- The actuation angle is relatively small because the cantilever is shorter than it is wide.
- The hinge-like suspension at the bottom of the plate does not only lead to a rotation but also to a large up and down movement. At large angles the filter can leave the beam.
- The power consumption of the parallel heater wires over the whole width of the plate is high, the force in parallel is not needed.
- The temperature profile of the actuator is not uniform horizontally. This leads to lateral strain that is not converted in actuation.

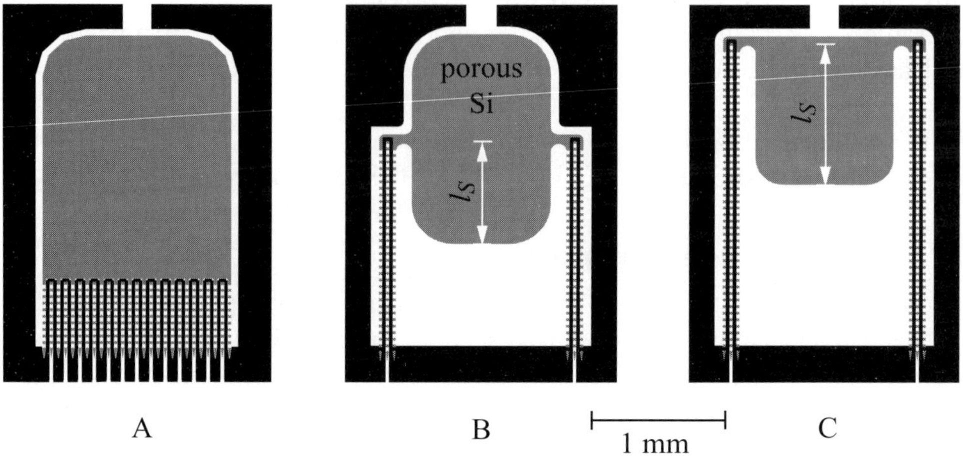

Fig. 7.13 Different designs to suspend the filter plate with actuators.
A: ribbon actuator at the bottom,
B: central arm suspension,
C: top arm suspension.
The metal layer is shown in black, the openings in the Si_3N_4 layer, where porous Si is formed, are shown in grey.

Another design uses two actuator arms to suspend the plate at the center of each side, Fig. 7.13 B. The actuation angle reaches 90° at a beam length of 1.6 mm. The length of the arms has to be at least twice as long as the distance l_S from the suspension point to the bottom edge of the filter plate to prevent the latter from touching the substrate when it stands up. This can be estimated as follows (see Fig. 7.14): Seen from the side, the circular shape of the arm

$$x^2 + (z - r)^2 = r^2 \tag{7.4}$$

can be approximated as quadratic function for small angles:

$$z = \frac{x^2}{2r} \tag{7.5}$$

The filter is flat and follows the slope of this function at the end point.

$$z' = \frac{x}{r} \tag{7.6}$$

It will cross the x-axis

$$z - l_S z' = \frac{x^2}{2r} - l_S\frac{x}{r} = 0 \tag{7.7}$$

at half the x-value of the tip

$$l_S = \frac{x}{2} \tag{7.8}$$

At higher angles the higher order terms of the circle function (Eq. 7.4) will result in additional height.

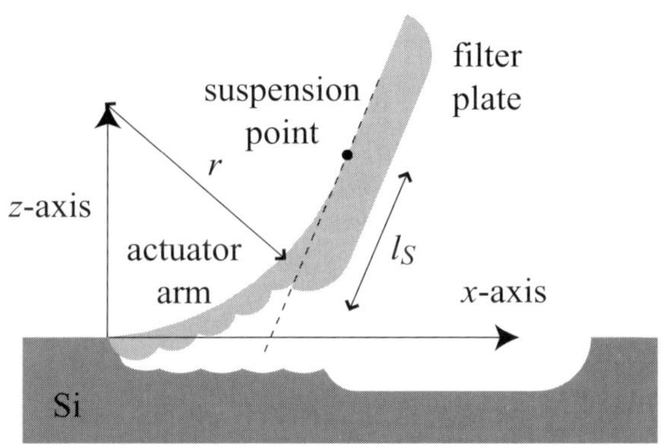

Fig. 7.14 Side view of optical filter plate (design of Fig. 7.13 B) standing up from the substrate. The length of the filter plate is limited by the arm length, otherwise it touches the substrate.

A third class of filters is suspended at the top edge, see Fig. 7.13 C. It has the longest arms with more than 2 mm, leading to high actuation angles. This design is also the most fragile of the three. The bondpads of all designs have been extended to the sides of the chip. This facilitates bonding so that the bond wires do not obturate the lightbeam.

The location where the structure broke first in early designs was the anchor point of the actuator with the substrate. To strengthen this part, the first perforation hole was modified from the square shape to a triangle to cause the current to flow out smoothly at the bottom (Fig. 7.13). This way the current density decreases gradually at the anchor point and inhibits electropolishing, linking the porous Si layer of the cantilever and the substrate. Further, the metal layer clamps the first two perforation holes of the actuator. At the suspension point of the plate, concave corners were rounded to prevent breaking (Fig. 7.13 B and C).

7.8 HOMOGENEITY OF THE OPTICAL FILTER

For optical filters it is important that the etch current density during porosification is equal over the filter area to get homogeneous layer thicknesses and a uniform optical wavelength. Krüger *et al.* reported that structured porous Si layers are thicker at the borders than in the center [253]. This effect is in the order of some percent for p^+ substrates up to a factor of two for p^- substrates. It results from a squeezing of field lines at the border of the structure (compare Fig. 7.2 b). On highly doped p^+ substrates the current density can equalize better, so that the inhomogeneity is less pronounced.

Krüger *et al.* proposed a metal mask to bias the etch voltage and create a uniform current distribution over the porosified and masked regions. This method – apart from being elaborate and needing design changes – could not be applied because it cancels all lateral currents and would inhibit detaching of the plate at its borders.

It was observed that the layer thickness also depends on the neighborhood of the porous Si structure. Mask openings with no other areas to porosify nearby lead to thicker layers than the same opening embedded in other structures. The reason for that effect is that all current lines from the unmasked backside of the wafer face only a small exit opening on the frontside, concentrating the current density.

Both problems, the variation of layer thickness within a mask structure and dependence on the neighborhood were reduced by a factor of about two with the following solution [240]: The wafer backside is masked with the same pattern as the frontside to equalize the current density. This limits the current arriving from the backside at the etch interface according to the size of the mask openings on the top and prevents current overshoot in areas with small structures like the actuators. For the fabrication process this means only few additional effort because Si_3N_4 is automatically deposited on front and backside in the LPCVD process and has to be etched away on the backside anyway. The additional step is a backside photolithography with the mirrored mask of layer 2. Resolution and alignment is not critical, it can also be done on an unpolished surface, it only serves to equalize the current densities.

To quantify the improvement of the backside mask, a finite element model (FEM) has been simulated. It assumes DC conductivity of the doped wafer and electrolyte. A 2D model was computed with the "FemLab" package based on MatLab. The Si p^+ wafer was modeled 500 µm thick, covered with an insulating Si_3N_4 mask with one 500 µm long and one 1500 µm long opening, spaced by 1000 µm, to simulate size and neighborhood effects. The electrolyte on both sides of the wafer was modeled using the specific resistance resulting from ion mobility. The specific resistance of HF / ethanol was measured with the double etch cell without wafer. It comes close to the value of HF in water solutions of the same concentration found in literature [128]. Fluid dynamics was neglected because the aim of the simulation was the current distribution in the wafer. Table 7.3 shows the specific conductivities used. The geometry was meshed in about 30000 triangles.

Table 7.3 Material conductivities for the FEM simulation of the etch current densities.

Material	Conductivity
Si p^+ doped	80 $(\Omega cm)^{-1}$
Si_3N_4	0 $(\Omega cm)^{-1}$
HF / ethanol	0.54 $(\Omega cm)^{-1}$
porous Si filled with HF / ethanol	0.54 $(\Omega cm)^{-1}$

As etching proceeds the geometry changes. The current distribution at the end of the process has been simulated with an additional 30 µm thick layer of porous Si at the mask openings. For the porous Si layer filled with electrolyte the conductivity of the latter was assumed.

7.8.1 Frontside mask only

First the etch current density with a frontside mask only has been simulated as a reference. The boundary condition is a fixed voltage source between the top and the bottom. The current field is plotted in Fig. 7.15. Higher current densities at the borders of the mask can be recognized. Note that for the optical layer thickness only the z-component j_z of the current vector j is relevant, corresponding to the etch rate in vertical direction (assuming linear etch rates, neglecting the change in porosity).

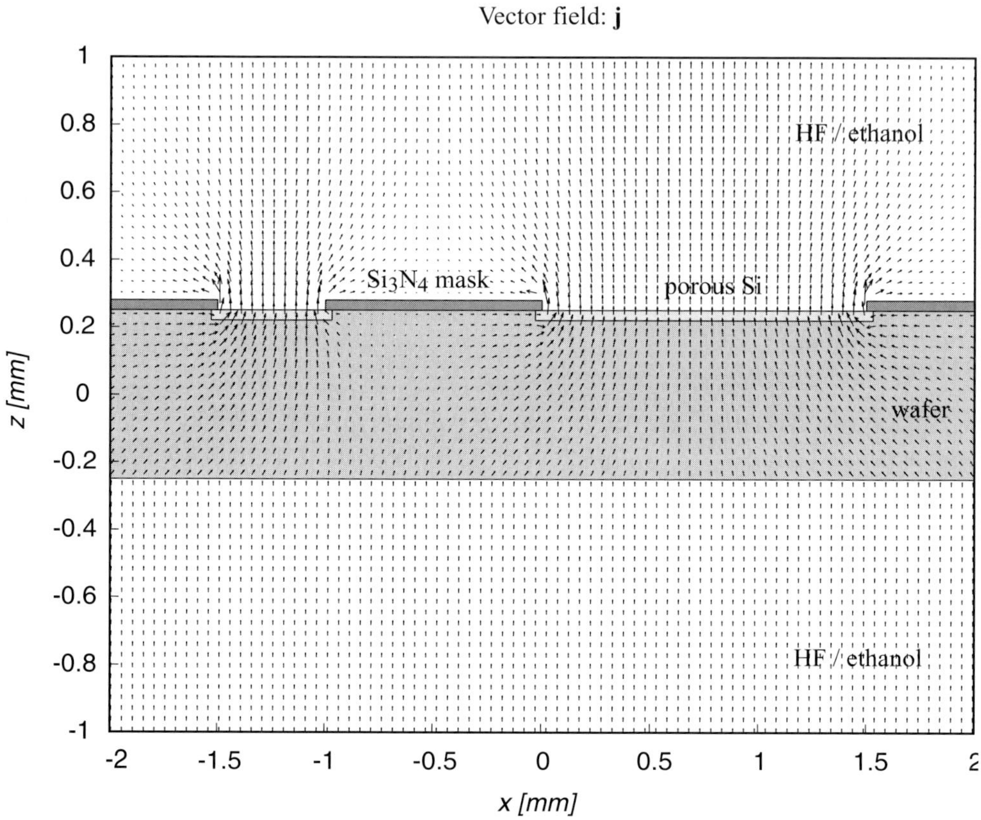

Fig. 7.15 Etch current field plotted as arrows for the case with a frontside mask only. The current density has an overshoot at the borders.

7.8.2 Frontside and backside mask

The second simulation shows the effect of the backside mask. The current field in Fig. 7.16 is more homogeneous. The overshoot at the borders has been reduced. The current density variations are most important directly at the surface and become less at a depth of 10 µm and more. Isotropic underetching creates new interface area where the current lines can spread out.

Fig. 7.17 plots j_z for different depths under the wafer surface. It can be seen that the inhomogeneities decrease with depth because the influence of the mask borders, where the current lines are concentrated, decreases. The simulations

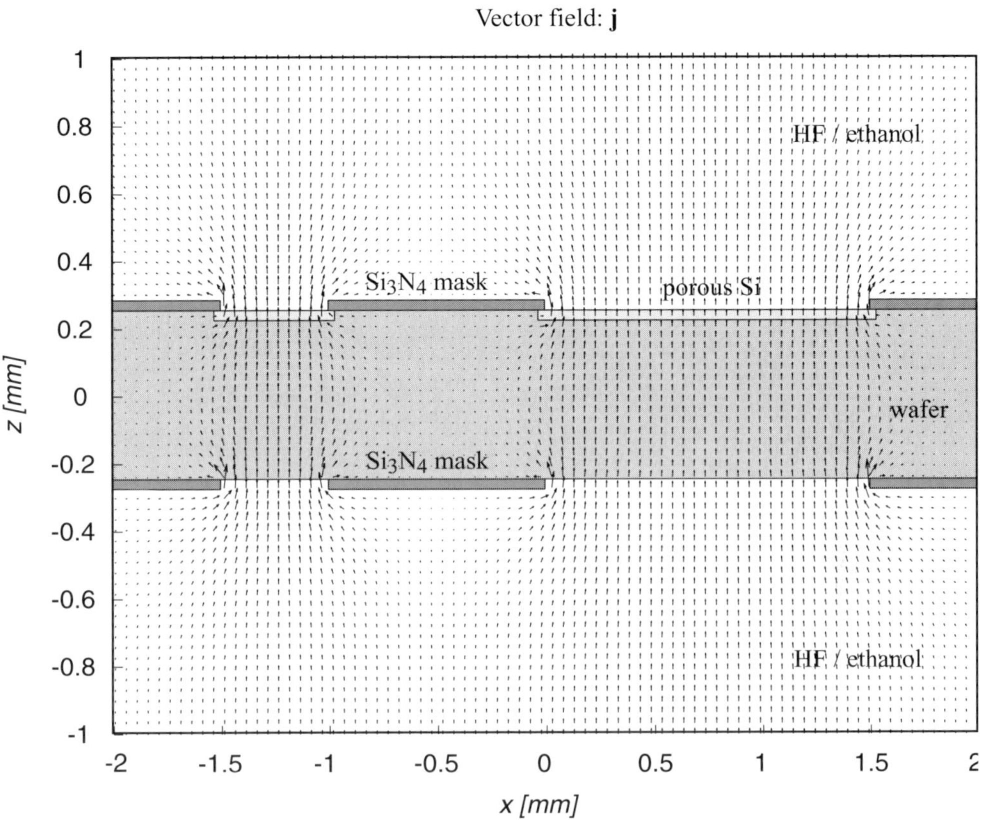

Fig. 7.16 Etch current field for the case with front- and backside mask. The current distribution becomes more homogeneous.

show that the backside mask improves not only the lateral homogeneity but also the depth homogeneity. Neighboring effects with other structures around are reduced.

The variation of the optical thickness of the layers scales only with an exponent of about 0.4 with the variations of the current density as can be seen from the calculations in section 6.8.3. This means that the filter wavelength homogeneity is better than the homogeneity of the current density.

7.8.3 Experimental comparison

The experimentally obtained inhomogeneities can be quantified with local wavelength measurements on the top and the bottom of the optical filter plate. The reflected central wavelength of a Bragg mirror has been measured at normal

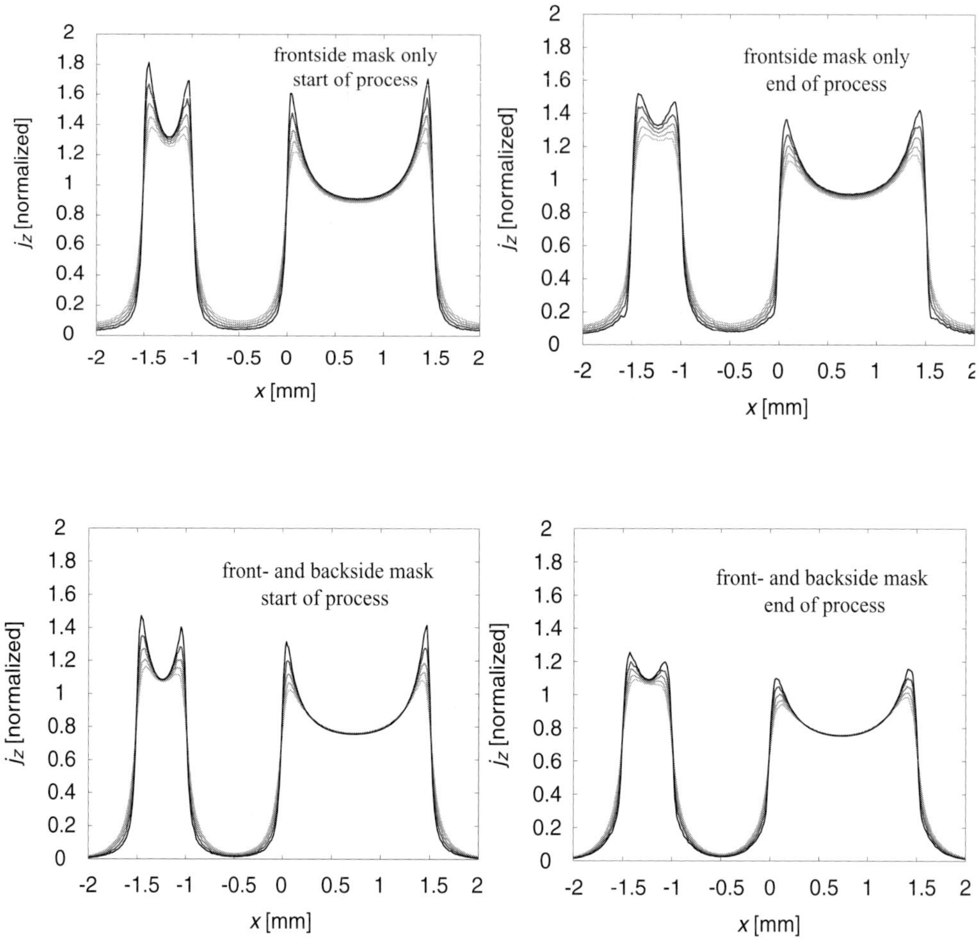

Fig. 7.17 Etch current density j_z for the case with and without backside mask. Cuts at different depths under the etch interface are shown from 30 μm (black) to 70 μm (light gray) in steps of 10 μm. The current density becomes more homogeneous at the end of the process because the current concentrated at the mask borders can distribute over the sidewalls of the porous layer.

incidence with a Nanospec 9100 spectrometer. Table 7.4 summarizes the deviation for filters fabricated with and without backside mask. Looking at the improvement, the backside mask is recommended.

Table 7.4 Experimental comparison of wavelength deviations between the center and the border of the filter for one device processed without and one with backside mask. The lateral variations can be reduced to one half with backside mask. The second filter was not fully compensated in depth homogeneity.

Filter	λ in the center	λ at the border	$\Delta\lambda$
Frontside mask only:			
upper side	620 nm	650 nm	30 nm
lower side	630 nm	640 nm	10 nm
With backside mask:			
upper side	600 nm	615 nm	15 nm
lower side	620 nm	630 nm	10 nm

7.8.4 Electropolished well

A method to improve the optical filter homogeneity even further is to electropolish additionally a well on the Si surface before porosification. All other

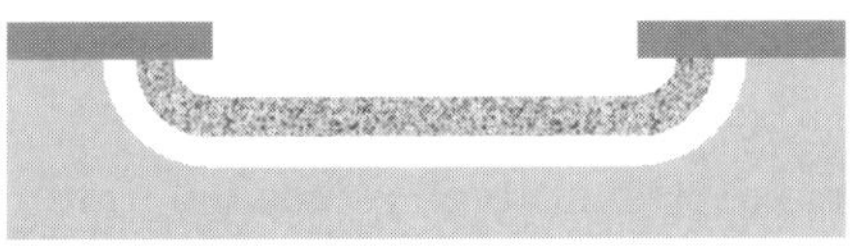

Fig. 7.18 Electropolishing before porosification generates a well. It improves the optical homogeneity and makes the filter plate mechanically more rigid. At the same time the actuator arms loose flexibility.

process steps remain, only the etch program is changed so that the open Si surface is etched isotropically 10 µm deep or more before porosification. Fig. 7.18 illustrates the cross section of the resulting plate with well.

As it was seen from the simulation, the etch current is more homogeneous at some depth than directly at the surface. Consequently the filter homogeneity improves when the first 10 µm are etched away completely and filter fabrication starts where the current lines are not concentrated any more at the border.

Mechanically this well makes the plate more rigid because it gets a rim at the border. MOEMS mirrors with rims have been proposed to improve optical flatness, especially against dynamic deformation [259, 260]. The electropolished well is a very elegant way to produce such a rim without any additional process steps. The drawback for this application is the fact that also the actuator becomes more rigid. The arm consists now of half-shells instead of half-spheres of porous Si. The actuation angle decreases by more than 50% for devices with electropolished wells. For this reason the concept will not be continued.

7.9 PROCESS FLOW

In this section the process flow for the fabrication of tunable optical filters of porous Si will be summarized. Electroplated Ni is used as metal layer. In an alternative process the metal layer can be evaporated.

The process starts on p^+ (Boron) doped Si wafers with a conductivity of 0.010 to 0.015 Ωcm, diameter 100 mm, crystal orientation [100], 525 ± 25 μm thick, single side polished, grown by the Czochralski method. The doping homogeneity within the wafer is important for the optical properties. Certain low quality wafers showed variations along the radius that led to color rings in optical filters.

The wafers are cleaned in a standard RCA procedure, a sequence of a bath of NH_4OH and H_2O_2, a bath of HF and a bath of HCl and H_2O_2. An HF dip removes native oxide just before loading in the LPCVD oven, where 200 nm of low-stress Si_3N_4 is deposited.

The wafers are dipped again in HF before evaporating 20 nm Cr (from an e-beam heated source) and 50 nm Au (from a Joule heated source) on the frontside as a seed layer for electroplating. Standard positive photoresist (PR) is spun on the frontside, pre-baked, illuminated with mask layer 1 and developed, see Fig. 7.19a. It defines the mold for electroplating Ni. In a standard galvanic $Ni(SO_3NH_2)_2$ (nickel sulfamate) bath a layer of 0.80 μm Ni is deposited at a current density of 10 mA/cm^2 within 3.5 minutes, Fig. 7.19b. The photoresist is then removed in a solvent. The Au layer that appears where no Ni was deposited is etched in a KI and I_2 solution in 30 sec. using a N_2 bubble bath to minimize underetching. The Cr layer under it is removed in a $Ce(NH_4)_2(NO_3)_6$ and CH_3COOH solution in 15 seconds, see Fig. 7.19c.

Photolithography layer 2 is used to structure the Si_3N_4 under it. For good photoresist adhesion the wafers have to be primed with HMDS (hexamethyl-disilazane). The same standard positive resist is spun on front- and backside. It is illuminated with mask layer 2 on top and the mirrored mask layer 2 on the back-

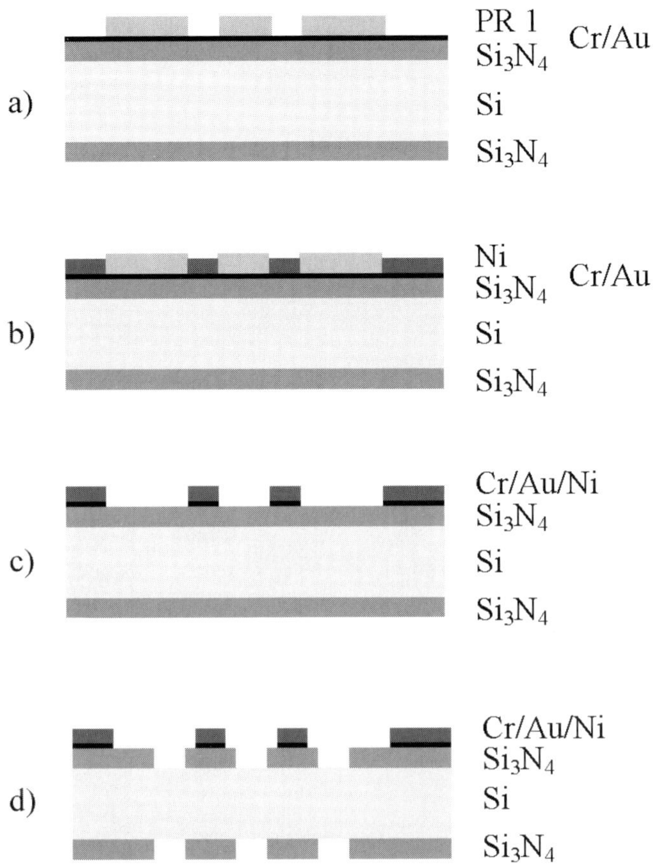

Fig. 7.19 Process flow for the fabrication of tunable optical microfilters, part 1. A cut through the actuator arm is shown (not at scale).

side and developed. Si_3N_4 on front and backside is structured with C_2F_6 reactive ion etching (RIE) using inductively coupled plasma (ICP), the result is shown in Fig. 7.19d.

To protect the Si surface for optical filters on the frontside, another layer of standard photoresist is spun on top. In order to prepare the wafer to be cleaved in chips at the end of the process, they are trenched on the backside 150 µm deep with a Ni saw blade 30 µm wide. The wafers are cleaned in acetone and isopropanole, removing the photoresist.

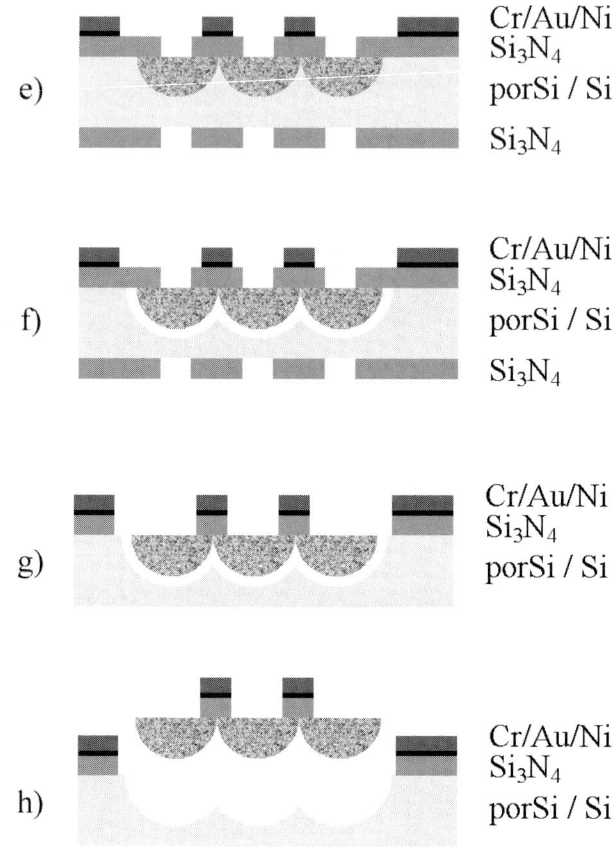

Fig. 7.20 Process flow for the fabrication of tunable optical microfilters, part 2. A cut through the actuator arm is shown (not at scale).

The porous Si layers are generated by anodic etching in HF / C_2H_5OH with the double cell wet etch processor described in section 6.3. A PC controlled constant current source applies current densities modulated between 50 .. 300 mA / cm^2 to obtain the multilayer interference filter and the actuator arms within 10 min, Fig. 7.20e.

The process step continues at a higher current density of 2000 mA / cm^2 for 2 min to etch 50 µm of Si isotropically by electropolishing, Fig. 7.20f. This step replaces the dissolution of a sacrificial layer in conventional surface micromachining.

Still in the same bath, the current is reversed and reduced to 10 mA / cm^2 for 30 min. This phase dissolves the Si_3N_4 mask where it is not covered by the metal layer, see Fig. 7.20g. At the same time the porous Si layer is conserved by the reversed potential that works similar to a cathodic anti-rust protection. Removing the Si_3N_4 mask releases the filter plate that was held down up to here at its underetched border. The structure flips up because of the internal strain between electroplated Ni and porous Si, Fig. 7.20h.

The wafer can be partly oxidized in oxygen-plasma in 10 min. This pre-ages the optical filter to prevent wavelength drift. Oxidation also increases the out of plane rest-angle. As shown in Fig. 7.10, the bottom part of the actuator arm is less porous. This means that during oxidation, the bottom part expands more due to the integration of oxygen atoms than the top [227], which finally increases the rest angle. The wafers are cleaved in chips at the predefined trenches. This will be discussed more in detail in chapter 8.

The fabrication process is of very moderate complexity compared to other tunable optical filters using MOEMS technology. It takes about one week for two to three people to carry out the complete cleanroom process as described.

7.10 Summary

Different insulating and conducting materials have been presented as masks for electrochemical etching of Si in HF. The selection is more restricted than for chemical HF etching because the applied potential can favor oxidation of metals, and gas bubbles can create adhesion problems for polymers. Low stress Si_3N_4 is most suited for the application envisaged. A new method has been developed to remove this mask without destroying the porous Si structure. The Si_3N_4 mask is combined with electroplated Ni wires on top, which serve as heater for a thermal actuator.

A thermal trimorph actuator has been presented, based on a new surface micromachining technology using porous Si. Design rules have been established. The actuator turns an optical filter, consisting of a porous Si multilayer. The actuation angle has been calculated in dependence of the dimensions. Different designs to suspend the plate were discussed.

The technology needs only two photolithography mask levels, one of which is partly perforated to etch different layer thicknesses and porosity gradients for the actuator and the filter plate. The current density distribution during electrochemical etching has been simulated to optimize the homogeneity of the optical filter. It has been shown that a backside mask can improve homogeneity by a factor of two. The technology and the device are patented.

8 Tunable Optical Filter and IR Gas Spectroscopy

8.1 Overview of Devices

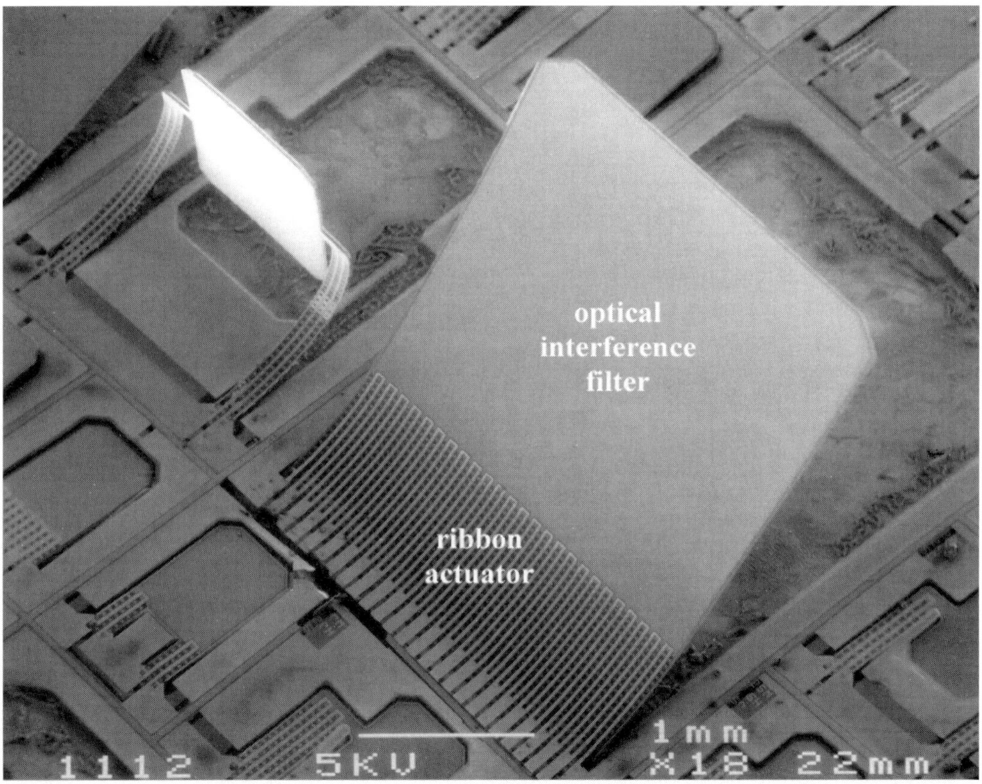

Fig. 8.1 SEM picture of a free-standing, tunable interference filter of density modulated porous Si lifted out of plane by a ribbon actuator. The plate contains 256 layers periods with high and low indices of refraction. Dimensions: 2400 x 3000 x 30 μm^3.

Large, flat, flip-up optical interference filters of porous Si were fabricated with areas ranging from 250 by 750 μm^2 to 2400 by 4000 μm^2 and a typical thickness of 30 µm. Fig. 8.1 shows the first design with a ribbon actuator (compare Fig. 7.13 A) . In the rest position, the plate is lifted about 45° out of plane. This angle can be increased with a longer actuator. Experiments with different thicknesses of the porous Si layer show that the plate begins to warp when it is thinner than 15 µm and is difficult to lift up when it is thicker than 40 µm, because the actuator gets also thicker and more rigid.

The filter in Fig. 8.2 is suspended by two arms, corresponding to the design B. This geometry leads to higher rest and actuation angles, but it is more fragile when large plates are required. For a filter size of 1100 x 1850 μm^2 it is the best compromise of actuation angle and stability. Most of the characteristics in the following refer to this design.

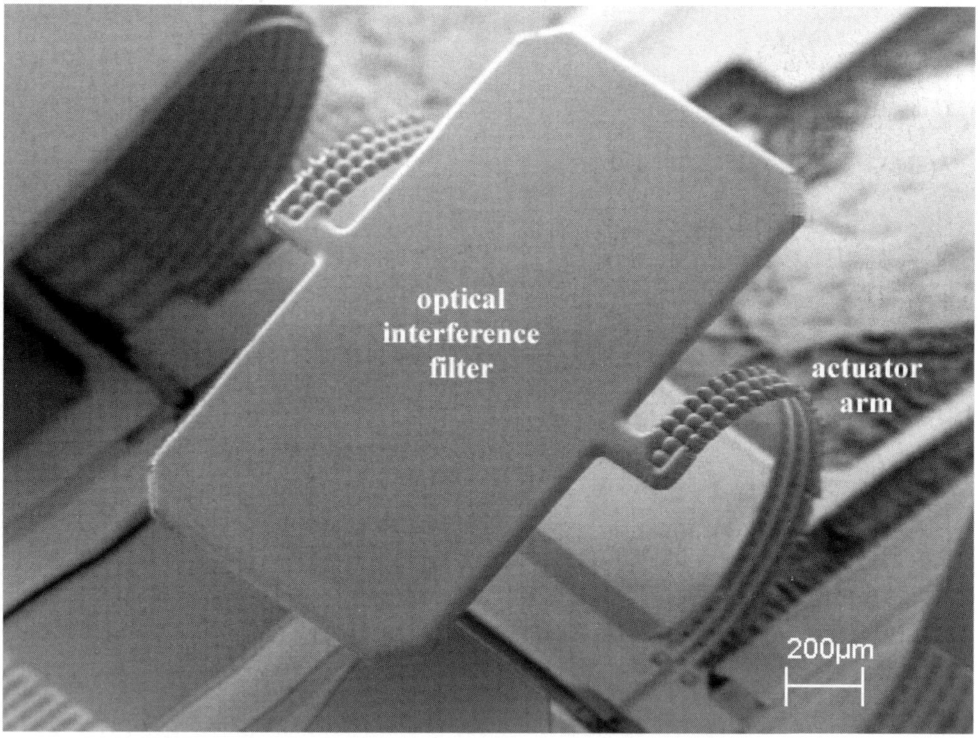

Fig. 8.2 SEM picture of a tunable interference filter with two actuator arms that lift it out of plane. Dimensions: 1100 x 1850 x 30 μm^3.

For tests on wafer level, the devices were contacted with prober needles. The actuator heater resistance is 30 Ω for the model shown in Fig. 8.2. The difference to the theoretical value of 22 Ω can be explained partly by the additional resistance of the bondpads and the contact resistance of the needles that contribute about 2 Ω. The rest can be explained by an increased specific resistance of the electroplated Ni layer compared to the assumed bulk value of 8.7 μΩ·cm. Specific conductivities ranging from 6.84 to 10.0 μΩ·cm can be found in literature for Ni of different qualities [128].

Fig. 8.3 shows a series of photos of a tunable filter plate taken at different DC actuation voltages (see also cover photo). Models for the resonance and cut-off frequency developed in chapter 2 can explain successfully the experimental data of different designs, see sections 4.3 and 4.4.7. Static and dynamic mechanical deflection of 90° of was achieved.

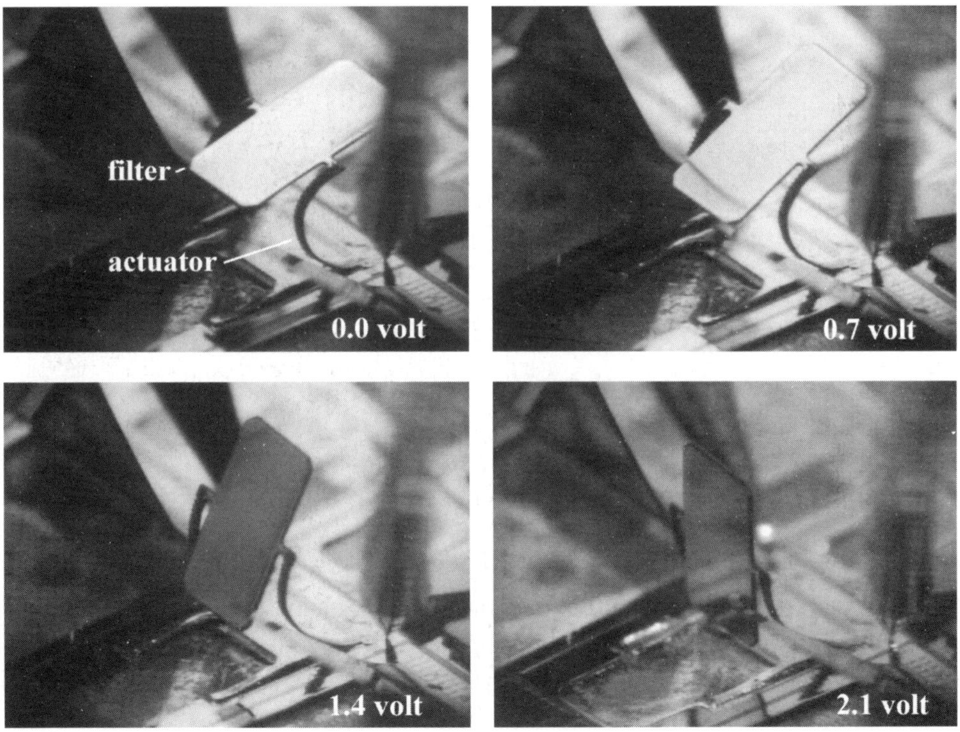

Fig. 8.3 Optical filter of porous silicon in motion. The tilt angle is a function of the actuator DC voltage and achieves 90°. The filter can also oscillate in resonance at 62 Hz. Dimensions: 1850 x 1100 x 30 μm^3.

8.2 Optical characterization

8.2.1 Visible light

The optical filter plate of (non-oxidized) porous Si is not transparent to visible light and can only be used as reflective filter. The angular dependence of the central wavelength was measured with a tunable Bragg mirror. The following experimental set-up was used: LEDs with colors from 592 nm (amber) to 635 nm (red) peak wavelength, 17 nm spectral halfwidth and 6° viewing angle pointed parallel to the chip substrate at the tunable Bragg mirror as illustrated in Fig. 8.4.

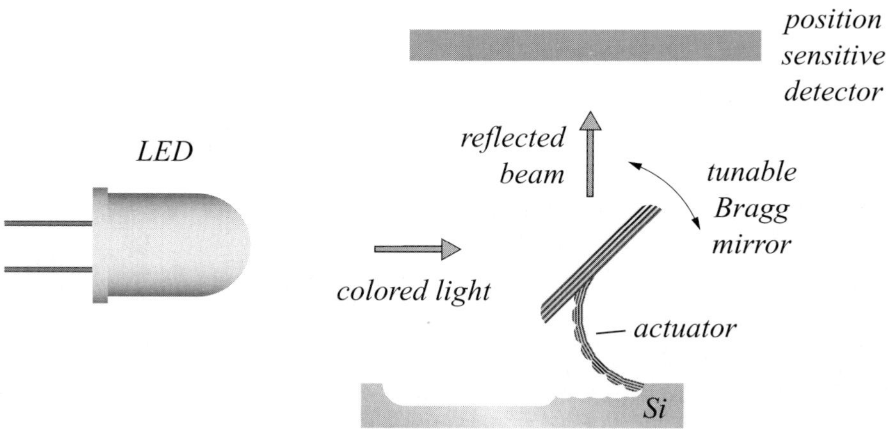

Fig. 8.4 Set-up for reflectivity measurement. The colored beam of a LED is reflected by a Bragg filter mirror. The mirror plate is turned to find the angle of maximum reflectivity for the specific wavelength.

The angle of reflection and the reflected intensity were simultaneously recorded with a position sensitive detector (PSD) during a sweep of the actuator angle, Fig. 8.5. The intensity was corrected for the projected area of the mirror and the different brightness' of the LEDs. The plot shows that wavelength differences of less than 20 nm can be resolved. This corresponds to the color resolution capabilities of the human eye in this wavelength region.

The peaks are smeared out because of thickness inhomogeneities of the porous Si filter plate due to current inhomogeneities during the electrochemical etch process discussed in section 7.8. The measurement has been taken on a device without backside mask.

For a microspectrometer, the detector does not need to be position sensitive, as the dependence of the filter angle on actuator voltage is known. A large area

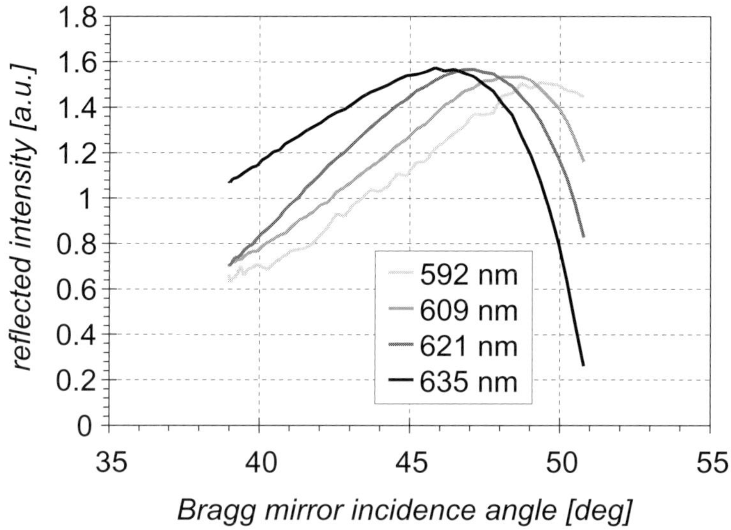

Fig. 8.5 Intensity reflected by a tunable Bragg mirror for different LED light sources. The wavelength can be selected by the incidence angle (equals the exit angle). Angular resolution: 1.5 ° (limited by beam divergence).

detector of some millimeters length covers the range of exit angles during the scan. However, it also collects background straylight. The more practical version is to fabricate a transmission filter that keeps the beam parallel to the substrate.

The porous Si filter can be made transparent to visible light by entire oxidation (porous SiO_2, see section 6.8.4). The filter spectrum shifts to the shorter wavelengths because the index of refraction decreases. This has to be taken into account in the optical design. Fig. 8.6 shows a tunable Bragg filter oxidized for 4 hours at 900° C in air.

Unfortunately, Ni heating tracks degrade during this oxidation so that the surface cannot be contacted electrically any more. Au tracks give better results after oxidation. Au also changes its surface properties and becomes granular, but it can still be used as electric heater after oxidation. The oxygen integrated in the porous Si of the actuator arms makes them stiffer, decreasing the maximum actuation angle considerably to about 10°. This limits the use of entirely oxidized tunable filters.

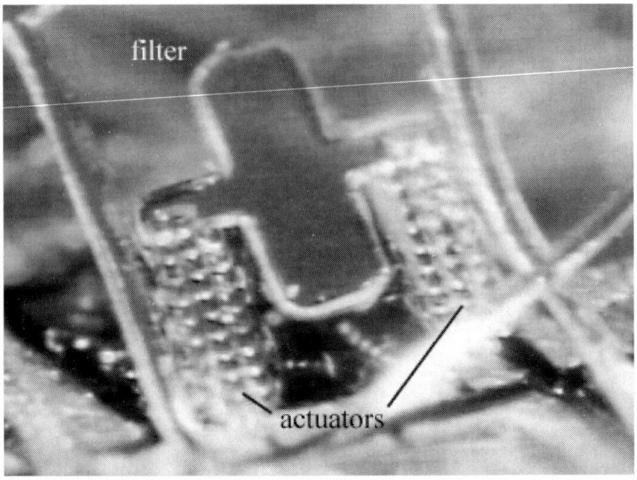

Fig. 8.6 Entirely oxidized tunable Bragg filter of porous SiO_2. Dimensions: 700 x 270 x 30 μm^3.

8.2.2 Infrared light

The main application for the tunable optical filter can be seen for infrared light. The possibility to build an IR spectrometer with only one single detector is the main advantage over existing expensive technology.

Infrared transmission filters of porous Si are easier to fabricate than visible light filters because Si is transparent beyond 1.1 μm, corresponding to its bandgap energy. The quantum confinement in nanoporous Si can shift the bandgap so that it reaches the visible spectrum. This does however not inhibit the infrared transparency. The mesoporous Si used for this device does not show a strong shift in bandgap energy. High doping in Si wafers can lead to absorption in the infrared spectrum. However, for our detached thin layers of less than 50 μm the effect can be neglected.

For spectroscopy the filter can be principally used in reflection or transmission. The transmission configuration is more favorable, the signal background is lower and the system is more compact. A Fabry-Pérot bandpass filter can be used to select a narrow band of the spectrum. However, their stop-bands are not very large, see section 6.7.4. Alternatively an edge filter can be used. It reflects and blocks out all wavelengths below the edge wavelength and transmits everything above. The detector signal during an angle sweep is then the integral of the spectrum of the beam to be analyzed from the edge wavelength onwards.

The filters were characterized with a Perkin Elmer 983 infrared spectrometer. It needs filter areas larger than 2 mm^2 to give reasonable signal to noise ratios. A large unstructured filter was used to measure the transmission spectra at different angles of incidence. It was fabricated on a p^+ Si wafer and detached from the substrate by electropolishing like the micromachined filters. A thicknesses of 30 μm is fragile but can be handled, 50 μm is more robust. The filter has been turned manually. Fig. 8.7 shows the IR transmission spectra of a Bragg filter for incident angles of 0° to 60°. The central wavelength of the blocking band is shifted according to Bragg's law (Eq. 6.21). A fit of the wavelength versus incidence angle gives an average index of refraction of 1.74, which agrees within a tolerance of 0.01 with the designed value.

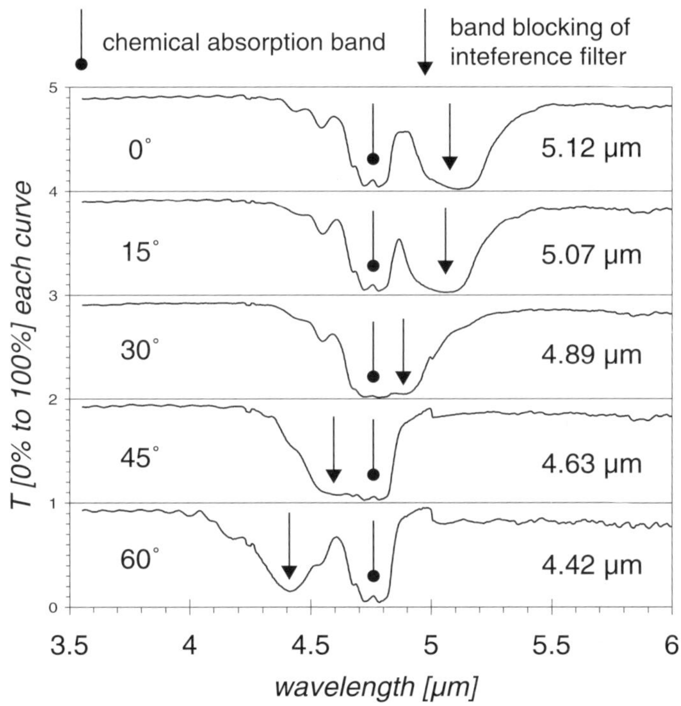

Fig. 8.7 Measured spectral transmission for different beam incidence angles. The center wavelength of the Bragg filter is tuned from 4.42 μm to 5.12 μm.

Additionally to the band that is blocked out by the interference filter, a constant band at 4.7 to 4.8 μm can be seen. A reflectivity measurement revealed that the band is absorbed. It can be identified as SiH, SiH_2 and SiH_3 stretching vibrations [258].

The central wavelength of the filter can be tuned by a factor of $\lambda_{max} / \lambda_{max} = 1.16$, which is sufficient to measure the absorption bands of different gases, as will be seen in the section 8.4.

8.3 CHIP SEPARATION AND PACKAGING

8.3.1 Cleaving

Separating the wafer with the filters released in an out-of-plane position is a delicate step. Sawing would damage the structures by the water jet and dust. The wafers are cleaved along separation trenches that have been sawed 150 μm deep on the backside with a 30 μm wide Ni blade before the last etching step.

Fig. 8.8 shows how the wafer is cleaved by lifting the support on one side with a knife-edge on top as counterforce. A special manipulator has been constructed for this purpose, see Fig. 8.9. When the wafer pieces become very small they can be sticked to adhesive blue tape as it is used for wafer sawing. This prevents loosing the chips.

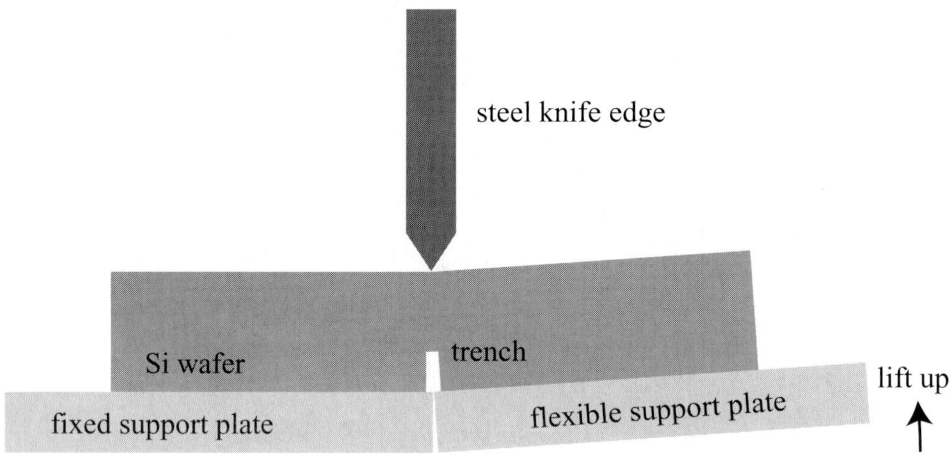

Fig. 8.8 Cleaving a wafer at a trench with a knife edge. The support plate is lifted manually (side view).

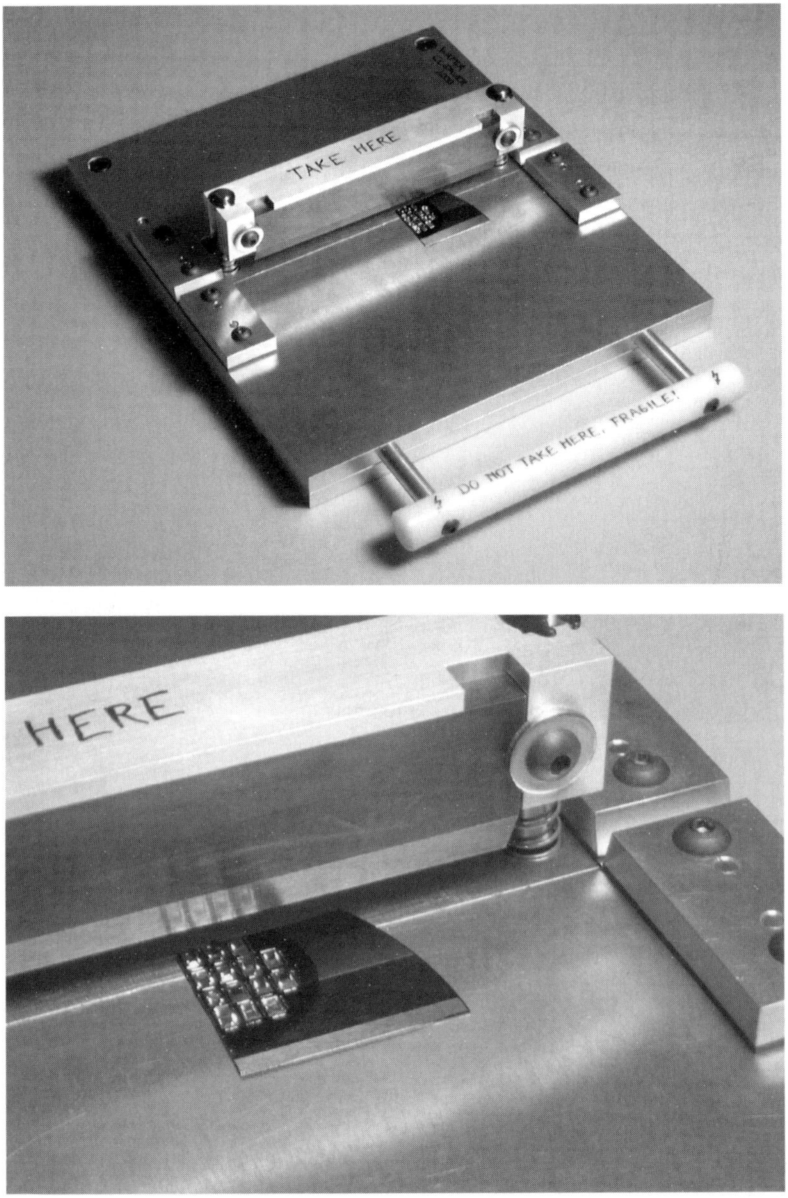

Fig. 8.9 *Wafer cleaver for 100 mm wafers. The knife edge in the center (marked "take here") is pressed on the pre-trenched cleave line on the wafer. The hinged front plate is lifted by a handle (white bar) to cleave the wafer. Dimensions 16 x 20 cm^2.*

8.3.2 Bonding on PCB

Microelectronic packages are generally not suitable for optical connections. The simplest solution is to glue the chip with the tunable optical filter on a printed circuit board (PCB) with epoxy glue. The Ni actuator can be connected electrically to Cu tracks on the PCB by standard ultrasonic Al wire bonding.

A detector is integrated on the PCB directly behind the filter. Fig. 8.10 shows how a thermopile (S707) in a TO-5 package is aligned in a slit milled in the PCB. The assembly precision needed is 20 to 50 μm, which is standard in electronics industry.

Fig. 8.10 Hybrid integration of a tunable optical infrared filter with a thermopile detector on a PCB.

8.3.3 Encapsulation

The filter plate is sensitive to air currents. To protect it also from touching, a plastic cover is glued on the PCB. An entrance window for the beam can consist of KBr, which is transparent for IR light over a large wavelength region, but the material is expensive and hygroscopic. Double sided polished Si wafers are cheaper and more robust, but about 30% of the intensity is lost at each interface because of the high refractive index.

8.3.4 Protection by a fusible link

The design can be improved by a fusible link on the metal layer at the top corners of the filter plate that holds it in a flat position after it is detached from the substrate until it is needed. It stays in the chip plane during drying, cleaving, placing the chip on a PCB, bonding and transport. The link is broken before the first use by a high current via a third bondpad or mechanically. This increases fabrication yields in the phase of system integration.

8.4 SYSTEM INTEGRATION FOR GAS SENSING

One of the most reliable methods to sense gases is measurement of their IR absorption bands. Every chemical compound has characteristic quantum mechanical vibration and rotation modes that correspond to specific energies and wavelengths that allow to identify and quantify it. The method is independent of humidity and temperature – the major difference with other detection methods like metal oxide gas sensors. Chemically not reactive molecules like CO_2 can be detected, too. Infrared absorption measurement has also a high selectivity for different combustible gases that can not be distinguished with catalytic gas detectors. No chemicals are used up like in electrochemical cells and an infrared sensor cannot be damaged by high gas concentrations.

When Max Planck found already in the year 1900 his famous radiation formula, he fitted the spectra of blackbody radiation. The data were measured with a rock salt prism and a bolometer wire as detector. The curves contained break-ins due to absorption by CO_2 and water vapor, see Fig. 8.11 [270].

8.4.1 Principle of infrared gas absorption spectroscopy

The absorption spectra of gases can be found in databases [271]. Fig. 8.12 shows the absorption cross section versus wavelength for 8 different gases. If a beam of wavelength λ and initial intensity $I_0(\lambda)$ passes a gas of the concentration c_{gas} (in particles per volume) over a path length l, the exit intensity $I(\lambda)$ will be given by

$$I(\lambda) = I_0(\lambda) \cdot \exp(-c_{gas} \cdot l \cdot \sigma(\lambda)) \tag{8.1}$$

where $\sigma(\lambda)$ is the absorption cross section area of the gas. For small concentrations the exponential function can be approximated linearly, which is known as the Beer-Lambert law.

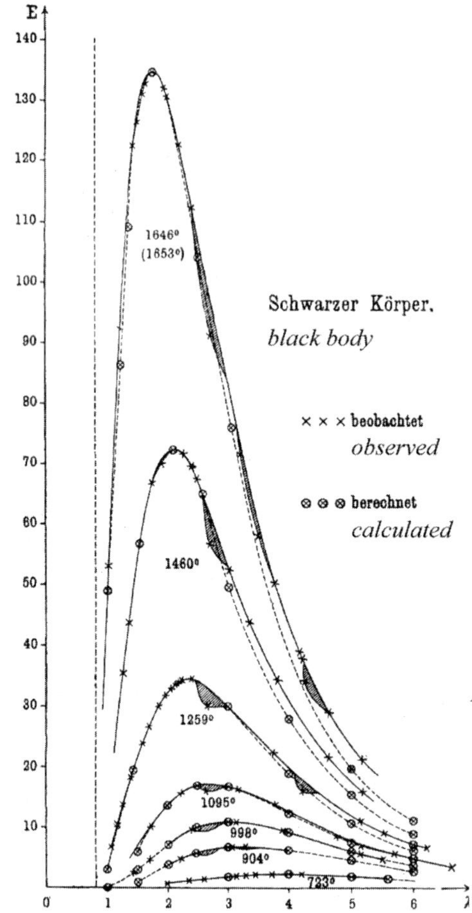

Fig. 8.11 Blackbody radiation spectra, measured around the year 1900. Systematic "errors" result from absorption by CO_2 at 4.3 µm and water vapor at 2.7 µm.

Standard infrared gas detectors use hot filaments as infrared source, followed by a gas cuvette of defined path length. The transmitted intensity at the absorption band is normally measured with a fixed bandpass filter, made for a specific gas, in front of a detector. The source has to be modulated because pyroelectric detectors give only an AC signal and drift with temperature, as do thermopiles. For that purpose a chopper wheel obturates the beam periodically, or the source is modulated electrically, if it is fast enough.

The measured intensity loss by the gas is used to calculate the concentration. The drawback is that the difference to the signal without gas is small and fluctuations in the source intensity can be erroneously interpreted as gas concentrations.

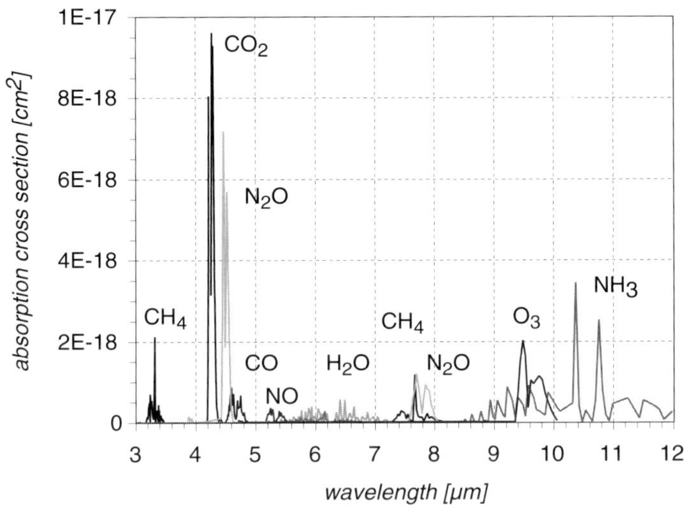

Fig. 8.12 Infrared absorption spectra of industrially important gases. The cross section can be regarded classically as the disk area of a molecule that a photon has to hit to get absorbed.

This effect can be reduced by a reference path at another wavelength. Double element infrared detectors with two different entrance filters for the gas absorption band and a reference wavelength are available for that purpose [272]. However, the parallel paths can be misaligned geometrically and differ in intensity without the presence of a gas.

The new concept presented here consists in having the absorption and reference wavelength in the same path with only one detector and tuning the interference filter in front of it. The intensity at the absorption wavelength and at the reference wavelength are measured serially in time. No chopper wheel is needed. Instead of modulating the intensity, the tunable filter modulates the wavelength. The signal to background ratio will be improved, as the optical signal and reference path are the same. Practical problems like straylight or the absorption by dirt on the windows of the gas cell are reduced, because they affect the reference and the gas absorption wavelength equally.

8.4.2 Set-up

A miniature light bulb with an IR transparent quartz glass housing is used as source. The beam is collimated with a parabolic mirror of metallized plastic. The spectral power density *dP/dk* of blackbody radiation is

$$\frac{dP}{dk} = \frac{2hc_0^2k^3A}{\exp(hc_0k/k_BT) - 1} \tag{8.2}$$

h Planck constant, 6.626e-34 Js,
c_0 speed of light, 2.998e8 m/s,
k_B Boltzmann constant, 1.381e-23 J/K,
k wavenumber, equals $1/\lambda$,
A surface area of blackbody,
T temperature of blackbody.

To get a maximum spectral density (per total emitted power) in the CO_2 absorption band at 4.26 µm (2350 cm^{-1}) the source temperature should be 572 °C, given by

$$T_{opt}(k) = \frac{hc_0k}{4k_B} \tag{8.3}$$

To achieve higher power with the limited filament area the lamp is operated at about 1500 °C, with a maximum spectral density per total power at 2.03 µm. The lamps consumes 500 mW.

The beam passes a 10 cm long cuvette flushed with the gas to analyze. It will be partly absorbed according to Eq. 8.1. Next it passes the tunable filter with the transmissivity $T(k,\alpha)$ that depends on the wavenumber k and the incidence angle. The total power arriving at the detector in dependence of the filter angle α and the concentrations c_1 .. c_n of n different gases with the absorption cross section areas $\sigma_1(k)$ .. $\sigma_n(k)$ is

$$P_{det}(\alpha, c_1 \ldots c_n) = \int_0^\infty \frac{dP}{dk} \cdot T(k, \alpha) \cdot \prod_{i=1}^{n} \exp(-c_i \cdot l \cdot \sigma_i(k)) \cdot dk \tag{8.4}$$

8.4.3 Experimental results

A Si based thermopile S707 (Dexter Research) with a responsivity of 153 V / W and a time constant of 12 ms was used as detector. Its sensitive area is

70 μm by 750 μm. A PC with DA/AD card controlled the angle of the filter and read out the preamplified detector voltage. The value was averaged over 1 sec, filtering out the 50 Hz power supply ripple. For a first proof of principle a macroscopic filter of porous Si with 22 mm diameter, turned with a motor, was used because it could be characterized with a Perkin Elmer IR spectrometer. Fig. 8.13 shows the detected transmission for 0.5 % CO_2 and 1% CO relative to N_2 as reference gas at filter incidence angles between 0° and 60°. It can be seen that the two gases are detected selectively by their different absorption bands.

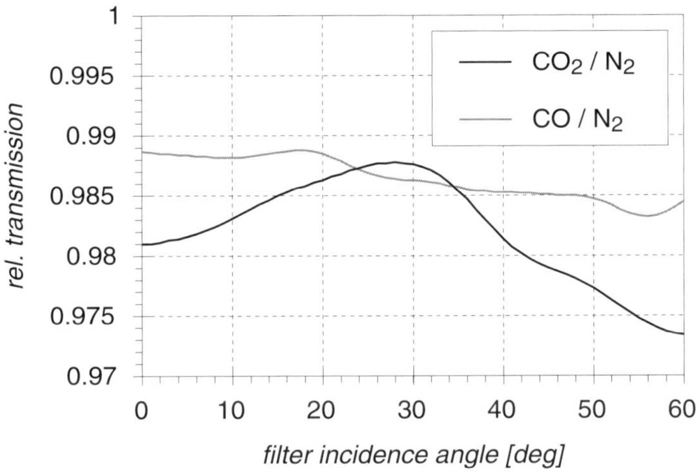

Fig. 8.13 Measured overall transmission (gas cell and filter) for different gases relative to a N_2 reference. The incidence angle of the filter was scanned from 0° to 60°, corresponding to a wavelength sweep like in Fig. 8.7. CO_2 (0.5 %) and CO (1.0 %) are detected selectively by their different infrared fingerprints.

To show the feasibility of signal treatment with low-cost analog electronics, a demonstrator of a CO_2 sensor has been built. It drives the actuator at 6.25 Hz to tune the filter between 4.0 μm and 4.26 μm and reads the detector intensity synchronously twice per cycle. The transmitted intensity for the upper and the lower filter position are stored in a "sample and hold" stage. Their difference is proportional to the CO_2 concentration and can be compared with a threshold level to trigger an alarm.

8.5 SUMMARY

The tunable optical filter for visible and infrared wavelengths can scan the desired part of the optical spectrum within milliseconds. A single pixel detector measures serially the intensity at selected wavelengths. This concept avoids expensive linear detectors as used for grating spectrometers. Usual gain variation, offset problems and others are eliminated.

The refractive index of the filter microplate is gradually modulated in depth to create a Bragg mirror, Fabry-Pérot bandpass or edge filter for a transmission wavelength between 400 nm and 6 µm. Thermal trimorph micro-actuators tilt the plate by up to 90°, changing the incidence angle of the beam to be analyzed. This tunes the wavelength transmitted to the detector by a factor of 1.16.

The filter area can be chosen between 0.27 x 0.70 mm^2 and 2.50 x 3.00 mm^2, the thickness is typically 30 µm. A comparison of different designs showed that the central actuator suspension gave the best results regarding large actuation angles (90°), mechanical stability (over 100 million cycles tested) and power consumption (typ. 50 mW). The filters show good flatness as they are etched out of one monocrystalline material.

The wafer with the released three-dimensional filter structures can be separated in chips at pre-sawed lines with a special wafer cleaver. The chips can be mounted and bonded on PCBs using a standard microelectronics bonder.

The spectral finesse $\Delta\lambda / \lambda$ of 25 is sufficient for a lot of sensor applications, e.g. detection of CO_2 and CO in combustion processes by their IR absorption bands. The feasibility of a miniaturized IR absorption gas sensor with reference measurement using wavelength modulation of the filter and only one detector has been presented. This opens up promising applications in infrared and visible light microoptics, such as low-cost microspectrometers for gas and liquid analysis. Online colorimetry and color correction, e .g. of desktop printers, can be envisaged.

9 CONCLUSIONS AND OUTLOOK

9.1 CONCLUSIONS

9.1.1 Micromirror

A new type of thermally actuated micromirror capable of one- and two-dimensional scanning has been presented. This work involves a complete study of the thermal bimorph actuator, an adapted inexpensive technological process development, analytical models and simulations for the dynamic behavior of the microscanner, characterization and testing of the devices and finally their evaluation for specific applications. An important feature of the developed microscanner is its simplicity, which makes it a cost effective component. Obtaining high fabrication yields and inexpensive devices is still a great challenge for MEMS components.

The thermal bimorph actuator was designed for optimized sensitivity and to meet performance specifications for the given applications. A compromise had to be found between materials with good thermal expansion characteristics and materials adapted to the chosen technological process. Bimorph actuator dimensions were designed for sufficient actuation angle, speed, robustness and adequate initial micromirror position. The calculations were verified by experimental measurements and were found to be in good agreement.

A theoretical thermal analysis was done for the microscanner. In free convection mode, the largest part of the generated heat in the actuator is lost by conduction into the silicon substrate (88%) and into the mirror (10%) (lost by convection by the mirror). At resonance actuation, the enhanced convection cooling at the mirror raise up to 36%. For a given applied power on the beam, the average temperature increases as the beam length increases and the beam thickness decreases. The optimal design is often a compromise between thermal efficiency and mechanical stability. The length of the beam influences also the actuator response time. A trade-off has to be found between fast response time on the one hand, and thermal efficiency on the other hand.

Bimorph actuator characterization showed for some material as chromium the necessity of an initial thermo-mechanical "burn-in" phase, after which the actuator sensitivity is stabilized. The mechanical and electrical characteristics of the conductive layer, depending on the layer morphology, are directly involved in the actuator reliability. In a static operation mode, the relatively low power consumption (10 mW) for large optical scan angles of 30° is a typical result.

The fabrication process is simple, based on original technological steps allowing the integration of the indispensable functions in maximum four mask levels. The same process allows to obtain one- and two-dimensional microscanners. Good fabrication yields of 95% were achieved using a dry release that opens doors to scalable mirror designs. Previous designs were limited by wet etch liberation. Special efforts were done on mirror flatness, resulting in reproducible and homogeneous curvatures.

Various one-dimensional microscanner types have been realized and characterized in a dynamic mode. The devices can be excited at their fundamental resonance frequency between 100 and 600 Hz, depending on their geometries. The parameters involved for the natural resonance frequency are given by a simple analytical model. In order to increase the accuracy of the design, FEM simulations were carried out, taking into account the initial curvature and the large deflections of the microscanner. The measurements were in good agreement with the simulations.

The scanners, besides showing high angular deflection and low power consumption, present a very smooth and regular movement, with good scan repeatability. The devices are capable of mechanical scan angles greater than 90° for frequencies varying from 100 to 600 Hz. Up to now, no other microscanner yields such performance regarding deflection angle. Even other microscanners actuated by thermal bimorph cantilevers [103, 73, 55] reach only 20° of maximum mechanical scan angle. Many scanning applications require optical scan amplitudes of typically 60°, which can be achieved by an average drive power as low as 1 mW. The lifetime of the devices is excellent. The good dynamic behaviour of the one-dimensional microscanners open the possibility to a new generation of scanning heads for a barcode reader. A compact scanning system was implemented and demonstrated.

A two-dimensional microscanner presenting an "L"-shaped beam was realized. A combined resonant and non-resonant movement can be excited simultaneously. It is capable of maximal mechanical scanning angles of 7° for 5 mW in the two scanning directions. The raster scanning frequency varies from 1000 Hz to 5000 Hz depending on the geometry. Successful image displaying is shown. Even though the resolution of the device remains quite low, it presents good potential for other scanning applications such as two-dimensional barcode readers.

9.1.2 Tunable optical filter

Based in the same actuation principle a tunable optical filter has been presented, where the metal mirror is replaced by an optical interference filter made of porous silicon. It uses the following MOEMS technology:

- A new surface micromachining technology that combines porous Si and electropolishing,
- a methodology for design and process calibration for optical filters of density modulated porous Si,
- a new concept for a microspectrometer that can serve for IR gas sensing, which has been experimentally verified.

The combination of porous Si and electropolishing under perforated mask geometries leads to a new surface micromachining technology. Si_3N_4 and electroplated Ni are the most suitable mask materials for HF electrochemical etching with the possibility introduced here to remove the Si_3N_4 mask after etching. The technology enables to integrate optical filters of porous Si with micromechanics and to design new tunable filter devices. It needs only two mask levels for photolithography and no sacrificial layers. The technology and the tunable filter have been patented.

The optical filter microplate is made of porous silicon layers of modulated density. The fabrication process for these filters has been calibrated to correct depth inhomogeneities and reduce lateral wavelength variations.

Thermal trimorph actuators suspend the filter plate and turn it in order to change the light incidence angle and tune the transmitted or reflected wavelength. The design is based on design rules that have been established for a perforated mask geometry and its underetching. The result is a thermal trimorph actuator with a new undulated shape. It was dimensioned and verified with analytical models and simulations. Tunable filters with areas between 0.27 x 0.70 mm^2 and 2.50 x 3.00 mm^2 have been achieved at a typical thickness of 30 µm.

The filter inclines by up to 90° in DC mode as well as at mechanical resonance. Resonance frequencies range from 24 Hz for the largest models to 3 kHz for small filters. The analytical model describes well the resonance frequency of the different models. The actuator consumes typically 50 mW.

The filter wavelength shifts with the light incidence angle. At normal incidence the filter wavelength is a factor of 1.16 higher than when it is deflected by an angle of 60°. The variation is smaller than for a diffraction grating because the light beam inside the filter is refracted to the normal. Oxidation of the filter lowers the refractive index and increases the tuning range. However, oxidation limits the actuator performance. The wavelength resolution of the tunable filter is 20 nm for visible light. For infrared applications the tunable filter chip has been integrated with a thermopile infrared detector for a microspectrometer system. The feasibility of an IR gas sensor based on the tunable optical filter as key component has been demonstrated.

Compared to other devices with a fixed grating and a detection line like the LIGA microspectrometer (C. Müller, [47]) or the one of CSEM Neuchâtel [50], the device presented here is about a factor of 2 smaller, its wavelength resolution is a factor of 2 inferior, the wavelength range covered is smaller and it is more sensitive to vibrations because of the mobile filter plate. The main advantage of the tunable optical filter is its capability to measure an infrared spectrum with only one single detector. For infrared applications the price of a linear detector exceeds the price of the LIGA fabricated grating by a factor of 1000.

For this reason, the technology presented here should not be seen in competition with conventional spectrometers as used in laboratory equipment, but as new opportunity for applications where up to now only one or a few wavelengths have been analyzed, like in infrared gas sensors.

If the device is compared to arrays of detectors with different fixed wavelength filters, the situation is similar. For visible light, such arrays can be fabricated on a silicon chip [40], but for infrared applications the number of elements is mostly limited to two because of costs. Here, the tunable filter becomes attractive. Arrays of lasers with different emission wavelengths (VCSELs) are subject of research of several groups [42], as well as tunable infrared semiconductor lasers. Their power output is still weak compared to hot filament sources and tunability is restricted. Materials like III-V and II-VI semiconductors and complex epitactic layer processes determine their cost.

Compared with a MEMS grating that can be turned to direct different wavelengths to a detector [43], the advantage of the tunable filter presented here is the suppression of higher harmonic orders and a sufficient filter area for infrared detectors.

A spectral filter with very high performance is the "Polychromator", a dynamic diffraction array where the optical phase of every line element can be controlled individually with electrostatic actuators [46]. The chip uses also a single detector and can potentially sense a large number of chemical substances by infrared absorption. The fabrication process for that device is more elaborate than for the one described here.

9.2 OUTLOOK

This book has given a thorough study of devices and technologies for thermally actuated micro-mirrors and filters. The model devices developed, i.e. one- and two-dimensional scanners and the integrated microspectrometer demonstrate the relevance of the choice of thermal bimorph actuators. At this stage of development, these model devices are not yet within specifications for industrial applications. But the study has shown that no fundamental technical reason opposes to reach this goal.

More generally, the combination and monolithic integration of actuators, mirrors and optical filters can potentially lead to other innovative MOEMS devices in new applications. We hope that the concepts and technologies developed in this book can contribute to the emergence of future MOEMS devices.

APPENDICES

A.1 COMPLEMENT TO THE CURVATURE CALCULATION DUE TO RESIDUAL STRESS

As seen in chapter 2, section 2.3, the strain (in *x*-direction) of the layer of a bent cantilever can be expressed by the sum of (Formula A.1.1):

- the mean strain ε^*_{layer} due to the process conditions.
- the $\Delta\varepsilon$ strain variation when the beam is released (which is the same in the two layers)
- the strain due to the bending moment and the flexural rigidity: y / r.

$$\varepsilon_{layer} = \varepsilon^*_{layer} + \Delta\varepsilon - \frac{y}{r} \qquad \text{(A.1.1)}$$

Note that the bending strain is negative because for $y > 0$ (Cr), the moment tends to decrease the strain in Cr layer. The linear elastic condition leads to the expression of the normal stress (in *x*-direction) in the layer by:

$$\sigma_{layer} = \varepsilon_{layer} \cdot \frac{E_{layer}}{1 - \nu_{layer}} = \varepsilon_{layer} \cdot E'_{layer} \qquad \text{(A.1.2)}$$

where E_{layer} and ν_{layer} are the elastic modulus and the Poisson's ratio, respectively. E'_{layer} is the biaxial elastic modulus.

The equilibrium in the released cantilever is reached when the total forces ΣF and the total bending moments ΣM are equal to zero (index Cr for chromium and index SiO_2 for silicon dioxide):

$$\sum F = \int_{-t_{SiO_2}}^{0} (\sigma_{SiO_2} \cdot dy) + \int_{0}^{t_{Cr}} (\sigma_{Cr} \cdot dy) = 0 \qquad \text{(A.1.3)}$$

$$\sum M = \int_{-t_{SiO_2}}^{0} (\sigma_{SiO_2} \cdot y \cdot dy) + \int_{0}^{t_{Cr}} (\sigma_{Cr} \cdot y \cdot dy) = 0 \quad \text{(A.1.4)}$$

Let us develop Equation A.1.3:

$$E'_{SiO_2}\int_{-t_{SiO_2}}^{0}\left(\left(\varepsilon^*_{SiO_2} + \Delta\varepsilon - \frac{y}{r}\right) \cdot dy\right) + E'_{Cr}\int_{0}^{t_{Cr}}\left(\left(\varepsilon^*_{Cr} + \Delta\varepsilon - \frac{y}{r}\right) \cdot dy\right) = 0 \quad \text{(A.1.5)}$$

$$E'_{SiO_2}\left(y \cdot \varepsilon^*_{SiO_2} + y\Delta\varepsilon - \frac{y^2}{2r}\right)\Bigg|_{-t_{SiO_2}}^{0} + E'_{Cr}\left(y \cdot \varepsilon^*_{Cr} + y\Delta\varepsilon - \frac{y^2}{2r}\right)\Bigg|_{0}^{t_{Cr}} = 0 \quad \text{(A.1.6)}$$

$$\begin{aligned} &E'_{SiO_2}\left(t_{SiO_2} \cdot (\varepsilon^*_{SiO_2} + \Delta\varepsilon) + \frac{t_{SiO_2}^2}{2r}\right) \\ &+ E'_{Cr}\left(t_{Cr} \cdot (\varepsilon^*_{Cr} + \Delta\varepsilon) - \frac{t_{Cr}^2}{2r}\right) = 0 \end{aligned} \quad \text{(A.1.7)}$$

$$\begin{aligned} &E'_{SiO_2}\varepsilon^*_{SiO_2}t_{SiO_2} + E'_{Cr}\varepsilon^*_{Cr}t_{Cr} + \Delta\varepsilon(E'_{SiO_2}t_{SiO_2} + E'_{Cr}t_{Cr}) \\ &+ \frac{1}{2r}(E'_{SiO_2}t^2_{SiO_2} - E'_{Cr}t^2_{Cr}) = 0 \end{aligned} \quad \text{(A.1.8)}$$

Let us define the following constants:

$$A_1 = E'_{SiO_2}\varepsilon^*_{SiO_2}t_{SiO_2} + E'_{Cr}\varepsilon^*_{Cr}t_{Cr} \quad \text{(A.1.9)}$$

$$A_2 = E'_{SiO_2}t_{SiO_2} + E'_{Cr}t_{Cr} \quad \text{(A.1.10)}$$

$$A_3 = E'_{SiO_2}t^2_{SiO_2} - E'_{Cr}t^2_{Cr} \quad \text{(A.1.11)}$$

and Equation A.1.8 can be written as follows:

$$A_1 + A_2 \cdot \Delta\varepsilon + \frac{A_3}{2r} = 0 \quad \text{(A.1.12)}$$

Let us develop Equation A.1.4:

$$E'_{SiO_2}\int_{-t_{SiO_2}}^{0}\left(\left(\varepsilon^*_{SiO_2} + \Delta\varepsilon - \frac{y}{r}\right) y\,dy\right) + E'_{Cr}\int_{0}^{t_{Cr}}\left(\left(\varepsilon^*_{Cr} + \Delta\varepsilon - \frac{y}{r}\right) y\,dy\right) = 0 \quad \text{(A.1.13)}$$

$$E'_{SiO_2}\left(\frac{y^2 \cdot \varepsilon^*_{SiO_2}}{2} + \frac{y^2 \Delta\varepsilon}{2} - \frac{y^3}{3r}\right)\Bigg|_{-t_{SiO_2}}^{0} + E'_{Cr}\left(\frac{y^2 \cdot \varepsilon^*_{Cr}}{2} + \frac{y^2 \Delta\varepsilon}{2} - \frac{y^3}{3r}\right)\Bigg|_{0}^{t_{Cr}} = 0 \qquad \text{(A.1.14)}$$

$$E'_{SiO_2} \cdot \left(-\left(\frac{t^2_{SiO_2}}{2} \cdot (\varepsilon^*_{SiO_2} + \Delta\varepsilon) + \frac{t^3_{SiO_2}}{3r}\right)\right) + E'_{Cr}\left(\frac{t^2_{Cr}}{2} \cdot (\varepsilon^*_{Cr} + \Delta\varepsilon) - \frac{t^3_{Cr}}{3r}\right) = 0 \qquad \text{(A.1.15)}$$

$$\frac{1}{2} \cdot (-E'_{SiO_2}\varepsilon^*_{SiO_2}t^2_{SiO_2} + E'_{Cr}\varepsilon^*_{Cr}t^2_{Cr}) + \frac{\Delta\varepsilon}{2} \cdot (-E'_{SiO_2}t^2_{SiO_2} + E'_{Cr}t^2_{Cr}) - \frac{1}{3r}(E'_{SiO_2}t^3_{SiO_2} + E'_{Cr}t^3_{Cr}) = 0 \qquad \text{(A.1.16)}$$

Let us define the following constants:

$$B_1 = -E'_{SiO_2}\varepsilon^*_{SiO_2}t^2_{SiO_2} + E'_{Cr}\varepsilon^*_{Cr}t^2_{Cr} \qquad \text{(A.1.17)}$$

$$B_2 = -E'_{SiO_2}t^2_{SiO_2} + E'_{Cr}t^2_{Cr} \qquad \text{(A.1.18)}$$

$$B_3 = E'_{SiO_2}t^3_{SiO_2} + E'_{Cr}t^3_{Cr} \qquad \text{(A.1.19)}$$

and Equation A.1.16 can be written as follows:

$$\frac{B_1}{2} + \frac{B_2 \cdot \Delta\varepsilon}{2} - \frac{B_3}{3r} = 0 \qquad \text{(A.1.20)}$$

From the solved system A.1.21 of two linear equation, $\Delta\varepsilon$ and r can be extracted (A.1.22 and A.1.23).

$$\begin{cases} A_1 + A_2 \cdot \Delta\varepsilon + \dfrac{A_3}{2r} = 0 \\ \dfrac{B_1}{2} + \dfrac{B_2 \cdot \Delta\varepsilon}{2} - \dfrac{B_3}{3r} = 0 \end{cases} \qquad \text{(A.1.21)}$$

$$\Delta\varepsilon = \frac{-4A_1B_3 - 3A_3B_1}{4A_2B_3 + 3A_3B_2} \tag{A.1.22}$$

$$\frac{1}{r} = \frac{6A_2B_1 - 6A_1B_2}{3A_3B_2 + 3A_2B_3} \tag{A.1.23}$$

If substituting the constants A_1, A_2, A_3, B_1, B_2 and B_3, after simplification A.1.23 can be written:

$$\frac{1}{r} = \frac{6 \cdot (\varepsilon^*_{Cr} - \varepsilon^*_{SiO_2}) \cdot t}{\dfrac{E'_{SiO_2} \cdot t_{SiO_2}{}^3}{E'_{Cr} \cdot t_{Cr}} + \dfrac{E'_{Cr} \cdot t_{Cr}{}^3}{E'_{SiO_2} \cdot t_{SiO_2}} + 6 \cdot t_{SiO_2} \cdot t_{Cr} + 4 \cdot t_{SiO_2}{}^2 + 4 \cdot t_{Cr}{}^2} \tag{A.1.24}$$

which is the Eqn. 2.9 of Chapter 2.

A.2 Complement to the Static Temperature Distribution Calculation

As seen in chapter 2, section 2.3, the static temperature distribution in the beam is expressed as follows:

$$T_b(x) = C_1 \cdot e^{\sqrt{\frac{2\alpha}{\lambda_b \cdot h_b}} \cdot x} + C_2 \cdot e^{-\sqrt{\frac{2\alpha}{\lambda_b \cdot h_b}} \cdot x} + C_3 \tag{A.2.1}$$

and the static temperature distribution in the mirror can be written:

$$T_m(x) = C_4 \cdot e^{\sqrt{\frac{2\alpha}{\lambda_m \cdot h_m}} \cdot x} + C_5 \cdot e^{-\sqrt{\frac{2\alpha}{\lambda_m \cdot h_m}} \cdot x} + C_6 \tag{A.2.2}$$

where $C_{1,2,3,4,5,6}$ are constants.

Let us define the following constants (for the meaning of the different parameters, see Chapter 2).

$$A_b = \sqrt{\frac{2\alpha}{\lambda_b h_b}} \tag{A.2.3}$$

$$A_m = \sqrt{\frac{2\alpha}{\lambda_m h_m}} \tag{A.2.4}$$

and

$$Z_b = \lambda_b h_b b_b A_b \tag{A.2.5}$$

$$Z_m = \lambda_m h_m b_m A_m \tag{A.2.6}$$

$$
\begin{aligned}
C_1 = & -\Big(q'_j Z_b\Big(e^{(A_m L_b)^2} + e^{(A_m L_t)^2}\Big) + q'_j Z_m\Big(e^{(A_m L_b)^2} - e^{(A_m L_t)^2}\Big) \\
& - q'_j Z_m(e^{(2A_m L_b + A_b L_b)} - e^{(2A_m L_t + A_b L_b)}) - 2T_{bulk}\alpha Z_b\Big(e^{(A_m L_b)^2} + e^{(A_m L_t)^2}\Big) \\
& - 2T_{bulk}\alpha Z_m\Big(e^{(A_m L_b)^2} - e^{(A_m L_t)^2}\Big) + 2T_{air}\alpha Z_b\Big(e^{(A_m L_b)^2} + e^{(A_m L_t)^2}\Big) \\
& - 2T_{air}\alpha Z_m\Big(e^{(A_m L_b)^2} - e^{(A_m L_t)^2}\Big)\Big) \cdot 1/\Big(2\alpha\Big(Z_b\Big(e^{(A_m L_b)^2} + e^{(A_m L_t)^2}\Big)\Big) \\
& + Z_b(e^{(2A_b L_b + 2A_m L_b)} + e^{(2A_b L_b + 2A_m L_t)}) + Z_m\Big(e^{(A_m L_b)^2} - e^{(A_m L_t)^2}\Big) \\
& + Z_m(e^{(2A_b L_b + 2A_m L_t)} - e^{(2A_b L_b + 2A_m L_b)})\Big)
\end{aligned} \tag{A.2.7}
$$

$$
\begin{aligned}
C_2 = & -\Big(q'_j Z_b(e^{(A_b L_b + 2A_m L_b)} + e^{(A_b L_b + 2A_m L_t)}) + q'_j Z_m\Big(e^{(A_m L_b)^2} - e^{(A_m L_t)^2}\Big) \\
& - q'_j Z_m(e^{(2A_m L_b + A_b L_b)} - e^{(2A_m L_t + A_b L_b)}) \\
& - 2T_{bulk}\alpha Z_b(e^{(A_b L_b + 2A_m L_b)} + e^{(A_b L_b + 2A_m L_t)}) \\
& + 2T_{bulk}\alpha Z_m(e^{(A_b L_b + 2A_m L_b)} - e^{(A_b L_b + 2A_m L_t)}) \\
& + 2T_{air}\alpha Z_b(e^{(A_b L_b + 2A_m L_b)} + e^{(A_b L_b + 2A_m L_t)}) \\
& - 2T_{air}\alpha Z_m(e^{(A_b L_b + 2A_m L_b)} - e^{(A_b L_b + 2A_m L_t)})\Big) \\
& \cdot (e^{A_b L_b})/\Big(2\alpha\Big(Z_b\Big(e^{(A_m L_b)^2} + e^{(A_m L_t)^2}\Big)\Big) \\
& + Z_b(e^{(2A_b L_b + 2A_m L_b)} + e^{(2A_b L_b + 2A_m L_t)}) + Z_m\Big(e^{(A_m L_b)^2} - e^{(A_m L_t)^2}\Big) \\
& + Z_m(e^{(2A_b L_b + 2A_m L_t)} - e^{(2A_b L_b + 2A_m L_b)})\Big)
\end{aligned} \tag{A.2.8}
$$

$$
C_3 = \frac{q'_j}{2\alpha} + T_{air} \tag{A.2.9}
$$

$$C_4 = Z_b e^{(A_m L_b)}\Big(q'_j e^{(A_b L_b)^2} - 2q'_j e^{(A_b L_b)} - 4T_{air}\alpha e^{(A_b L_b)} +$$

$$4T_{bulk}\alpha e^{(A_b L_b)} + q'_j\Big)$$

$$\cdot 1/\Big(2\alpha\Big(Z_b\Big(e^{(A_m L_b)^2} + e^{(A_m L_t)^2}\Big)\Big) \qquad \text{(A.2.10)}$$

$$+ Z_b(e^{(2A_b L_b + 2A_m L_b)} + e^{(2A_b L_b + 2A_m L_t)}) + Z_m\Big(e^{(A_m L_b)^2} - e^{(A_m L_t)^2}\Big)$$

$$+ Z_m(e^{(2A_b L_b + 2A_m L_t)} - e^{(2A_b L_b + 2A_m L_b)})\Big)$$

$$C_5 = Z_b\Big(q'_j e^{(A_b L_b)^2} - 2q'_j e^{(A_b L_b)} - 4T_{air}\alpha e^{(A_b L_b)} +$$

$$4T_{bulk}\alpha e^{(A_b L_b)} + q'_j\Big)$$

$$\cdot 1/\Big(2\alpha\Big(Z_b\Big(e^{(A_m L_b)^2} + e^{(A_m L_t)^2}\Big)\Big) \qquad \text{(A.2.11)}$$

$$+ Z_b(e^{(2A_b L_b + 2A_m L_b)} + e^{(2A_b L_b + 2A_m L_t)}) + Z_m\Big(e^{(A_m L_b)^2} - e^{(A_m L_t)^2}\Big)$$

$$+ Z_m(e^{(2A_b L_b + 2A_m L_t)} - e^{(2A_b L_b + 2A_m L_b)})\Big)$$

$$C_6 = T_{air} \qquad \text{(A.2.12)}$$

A.3 LARGE DEFLECTIONS

The beam curvatures, as well as θ or y discussed in Chapter 4 (Eqn. 4.2 and 4.3) and 5, were obtained by solving the differential equations for a beam with small angle of rotation (Eqn. A.3.1).

$$\frac{1}{r} = \frac{d\theta}{dx} = \frac{d^2 v}{dx^2} = -\frac{M}{E \cdot I} \tag{A.3.1}$$

where v is the bending deflection, M is the bending moment, EI the flexural rigidity of the beam (Fig. A.3.1 a).

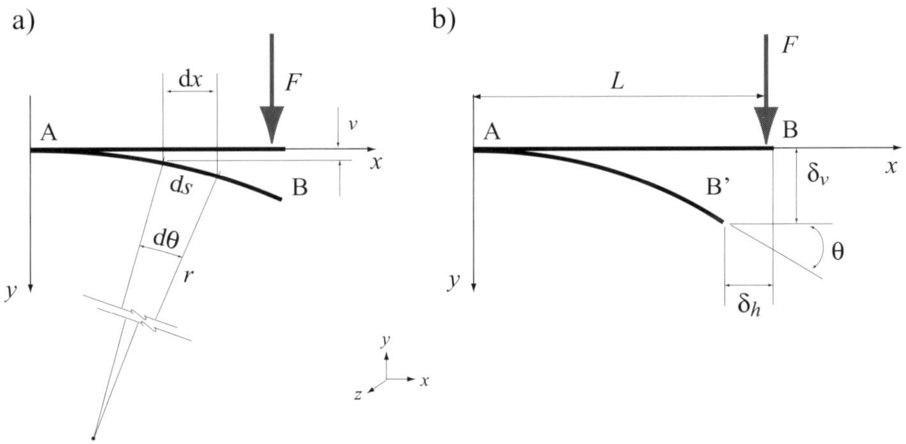

Fig. A.3.1 a) Deflection curve of a beam. b) Large deflection of a cantilever beam.

When slopes and deflections become large, the exact differential equation of the deflection curve must be used. This equation, which is also based upon the assumption that the material of the beam remains linearly elastic, is:

$$\frac{1}{r} = \frac{d\theta}{ds} = \frac{\frac{d^2v}{dx^2}}{\left[1 + \left(\frac{dv}{dx}\right)^2\right]^{\frac{3}{2}}} = -\frac{M}{E \cdot I} \tag{A.3.2}$$

where s is the distance measured along the curve itself. When the rotations are very small, the distance s becomes the same as the distance x, and the angle of rotation θ becomes the same as the slope dv/dx. However, for large deflections these simplifications are not valid. The exact shape of the elastic deflection curve, as found from Eqn. A.3.2, is called the elastica.

The mathematical solution of the problem of the elastica has been obtained for many different types of beams and loading conditions. Because the solutions are lengthy, they will not be worked out here. Instead, we will give the final results for one beam of practical interest. This allows us to estimate the error induced when considering small angle of rotation.

Let us consider the cantilever AB shown in Figure A.3.1 b. The load F is assumed to produce large deflections of the beam, resulting in the end of the beam moving from B to B'. The angle of rotation of the end of the beam is denoted θ, and the horizontal and vertical displacements are δ_h and δ_v, respectively. The length AB' of the deflection curve is equal to the initial length L because axial changes in length due to direct tension is disregarded. Because the beam is statically determinate, the expression for the bending moment M can be obtained easily and substituted into the differential Equation A.3.2. Then after considerable manipulation of the equation, the solution can be obtained in terms of elliptic functions. From this solution, expression for θ, δ_v and δ_h are obtained in terms of elliptic integrals [257]. The results in this case for θ, δ_h and δ_v are shown grahically in Figure A.3.2 a, b and c, respectively.

In each instance, the large deflection results merge with those obtained by small-deflection theory as the load approaches to zero. As the load get higher, the values of the small deflection theory are superior than the results for the large deflections. From Graph A.3.2 a, we can estimate the error for the angle θ between large deflection analysis and small deflection theory. The error to the exact value (large beam deflection) is *equal or greater than 10% for a mechanical angle equal or superior than 18º*. In the same way (Fig. A.3.2 b), for a typical beam length of 300 μm and 120 μm, above a vertical deflection δ_v of approximately 60 μm, and 12 μm, respectively, the error is more than 10%.

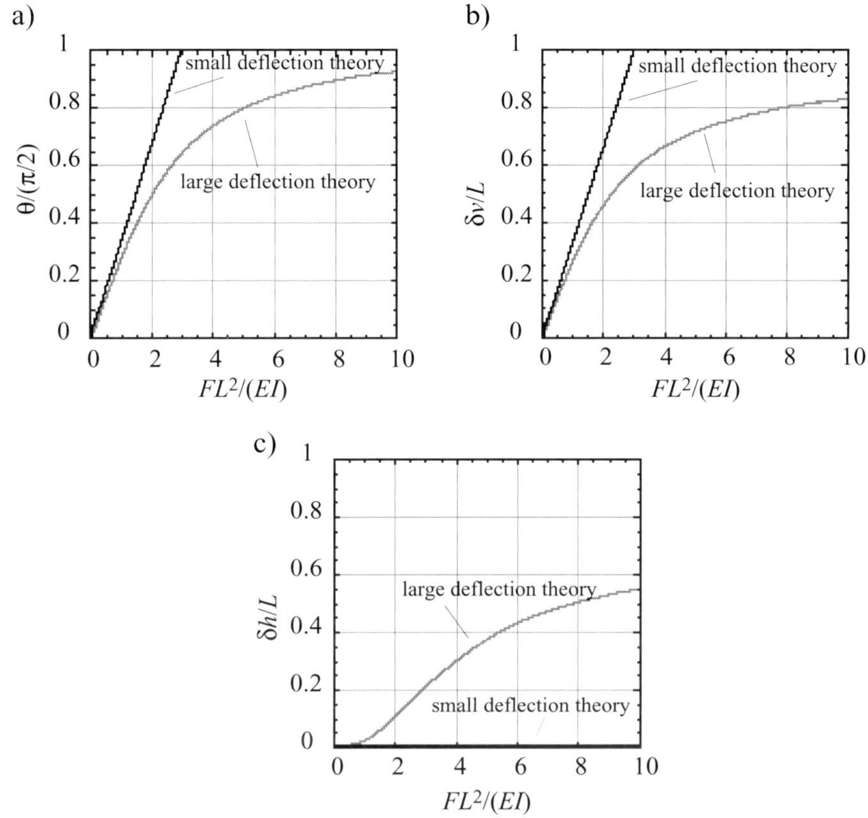

Fig. A.3.2 Large deflection of a cantilever beam with a concentrated load at the free end. a) angle of rotation θ, b) vertical deflection δ_v, and c) horizontal deflection δ_h. Adapted from Gere [257].

From the graphs, we can also estimate the equivalent force F at the tip of the cantilever due to a certain angle of deflection. Thus, for a mechanical angle of 20° and a beam length of 300 μm, we can calculate (large deflection theory) a force F of about 10 μN. For the same angle, a force of 60 μN is achieved in the case of a typical 120 μm long beam.

The complete solution for a cantilever beam with a vertical load at the end (Fig. A.3.1 b) is given in [273] and [274] and the deflection of a uniformly loaded beam is given in [275].

Finally, our model developed at the beginning of the chapter can predict rotations angles with an error less than 10% for optical angles lower than 36°.

REFERENCES

[1] R. W. Corrigan, "Scanned linear architecture improves laser projectors", *Laser Focus World*, pp. 169-172, 1999.

[2] K. E. Petersen, "Micromechanical light modulator array fabricated on silicon", *Applied Physics Letters*, vol. 31, pp. 521-523, 1977.

[3] L. J. Hornbeck, "Digital light processing and MEMS: timely convergence for a bright future", presented at *Micromachining and Microfabrication '95*, Dallas, Texas, USA, pp. 3-21, 1995.

[4] J. Bühler, "Deformable micromirror arrays by CMOS technology", Ph.D. thesis at the Swiss Federal Institute of Technology Zürich, Physical Electronics Laboratory, Diss. ETH no. 12139, 1997.

[5] A. Heuberger, B. Wagner, U. Hofmann, and K. Reimer, "Spatial Light Modulators and Scanners (1D & 2D)", Fraunhofer Institut für Siliziumtechnologie (ISIT), Itzehoe, Germany, 1998.

[6] W. Dötzel, Th. Gessner, R. Hahn, Ch. Kaufmann, K. Kehr, S. Kurth, J. Mehner, "Silicon mirrors and micromirrors arrays for spatial laser beam modulation", *Proceedings of Transducers '97*, Chicago, USA, pp. 81-84, 1997.

[7] M.-H. Kiang, O. Solgaard, and K. Y. Lau, "Electrostatic combdrive-actuated micromirrors for laser-beam scanning and positioning," *Journal of Microelectromechanical Systems*, vol. 7 (1), pp. 27-37, 1998.

[8] R. A. Conant, P. M. Hagelin, U. Krishnamoorthy, O. Solgaard, K. Y. Lau, R. S. Muller, "A raster-scanning full-motion video display using polysilicon micromachined mirrors", *Proceedings of Transducers '99*, Sendai, Japan, pp. 376-379, 1999.

[9] Ph. Hill, "Beam deflectors will be smaller and faster", *Opto & Laser Europe*, 1999.

[10] Th. A. Furness III, "Where there is no vision ...", *Proceedings of Eurosensors XII*, Southampton, UK, pp. 30-49, 1998.

[11] K. Y. Lau, "Microscanner Raster-Scanning Display: A spyglass for the future", *Optics & Photonics News*, vol. 5, pp. 47-50, 1999.

[12] K. Yamada and T. Kuriyuma, "A novel asymmetric silicon micro-mirror for optical beam scanning display", *Proceedings of MEMS '98,* Heidelberg, Germany, pp. 110-115, 1998.

[13] R. A. Miller and Y.-C. Tai, "Micromachined electromagnetic scanning mirrors", *Optical Engineering*, vol. 36, pp. 1399-1407, 1997.

[14] H. Kück, W. Doleschal, A. Gehner, W. Grundke, R. Melcher, J. Paufler, R. Seltmann, and G. Zimmer, "Deformable micromirror devices as phase modulating high resolution light valves", *Proceedings of Transducers '95, Eurosensors IX*, Stockholm, Sweden, pp. 301-304, 1995.

[15] G. Vdovin, S. Middelhoek, and P. M. Sarro, "Technology and applications of micromachined silicon adaptive mirrors", *Optical Engineering*, vol. 36, pp. 1382-1390, 1997.

[16] M. E. Motamedi, S. Park, A. Wang, M. S. Dadkhah, A. P. Andrews, H. O. Marcy, M. Khoshnevisan, and A. E. Chiou, "MOEM scan engine for bar code reading and factory automation", presented at LASE '98, Miniaturized Systems with Micro-Optics and Micromechanics III, San Jose, California, pp. 1346-1353, 1998.

[17] M.-H. Kiang, O. Solgaard, R. S. Muller, and K. Y. Lau, "Surface-micromachined electrostatic-comb driven scanning micromirrors for barcode scanners", *Proceedings of MEMS '96*, San Diego, CA, USA, pp. 192-197, 1996.

[18] M. Ikeda, H. Goto, M. Sakata, and S. Wakabayashi, "Two dimensional silicon micromachined optical scanner integrated with photo detector", *SPIE*, vol. 2383, pp. 118-124, 1995.

[19] K. Petroz, E. Ollier, H. Grateau, and P. Mottier, "Integrated silica micro-opto-electromechanical steering device for laser beam scanning", *Sensors & Actuators A*, vol. 73, pp. 117-121, 1999.

[20] K. Y. Lau, "MEM's, the word for optical beam manipulation", *Circuits & Devices, IEEE*, vol. 13, pp. 11-17, 1997.

[21] D. A. Roberts, R. R. A. Syms, A. S. Holmes, and E. M. Yeatman, "Dual numerical aperture confocal operation of moving fibre barcode reader", *Electronics Letters*, vol. 35, pp. 1659-1658, 1999.

[22] A. Wicki, V. Marsico, K. Kuhnke, K. Kern, L. Paratte, S. Schweizer, and Ph. Renaud, "A micromechanical detector for molecular beams", *Rev. Sci. Instr.*, vol. 70, pp. 3562-3565, 1999.

[23] H. Toshiyoshi, H. Fujita, "Electrostatic micro torsion mirrors for an optical switching matrix", *Journal of Microelectromechanical Systems*, vol. 5 (4), pp. 231-237, 1996.

[24] S.-S. Lee, L.-S. Huag, C.-J. Kim, M. C. Wu, "2x2 MEMS fiber optic switches with silicon sub-mount for low cost packaging", *Solid-State Sensor and Actuator Workshop*, Hilton Head Island, South Carolina, pp. 281-284, 1998.

[25] C. Marxer, C. Thio, M.-A. Gretillat, O. Anthamatten, R. Baettig, B. Valk, P. Vogel and N. F. de Rooij, "Vertical mirrors fabricated by deep reactive ion etching for fiber optic switching applications", *Journal of Microelectromechanical Systems*, vol. 6 (3), pp. 277-285, 1997.

[26] A. Ruzzu, D. Haller, J. Mohr, U. Wallrabe, "Optical 2x2 switch matrix with electromechanical microactuators", *MOEMS and Miniaturzied Systems, Proceedings of SPIE*, vol. 4178, pp. 67-76, 2000.

[27] V. A. Aksyuk, F. Pardo, C. A. Bolle, S. Arney, C. R. Giles, D. J. Bishop, "Lucent Microstar^{TM} micromirror array technology for large optical crossconnects", *MOEMS and Miniaturzied Systems, Proceedings of SPIE*, vol. 4178, pp. 320-324, 2000.

[28] M. Neumeier, W. Ehrfeld, J. Jäger, A. Picard, J. Schulze, "Miniaturized fiber optical switches with nonmoving polymeric mirrors for tele- and data-communication networks fabricated using the LIGA technology", *Miniaturized Systems with Micro-Optics and Micromechanics III, Proceedings of SPIE*, vol. 3276, pp. 37-47, 1998.

[29] http://www.agilent.com/about/feature/photonic.html

[30] S. C. Gustafson and T. A. Tuthill, "Micromirror arrays for active optical aberration control", *SPIE*, vol. 2687, pp. 8-17, 1996.

[31] H. Goto, S. Wakabayashi, M. Ikeda, M. Sakata, K. Imanaka, M. Takeuchi, and T. Yada, "Micro focusing optical device using piezoelectric thin film actuator", *SPIE*, vol. 2383, pp. 136-143, 1995.

[32] M. Strassner, J. Daleiden, N. Chitica, D. Keiper, B. Stålnacke, S. Greek, K. Hjort, "III-V semiconductor material for tunable Fabry-Perot filters for coarse and dense WDM systems", *Sensors & Actuators A*, vol. 85, pp. 249-255, 2000.

[33] N. Chitica, J. Daleiden, M. Strassner, K. Streubel, "Monolithic tuneable InP-based filter with 10 nm bandwidth for optical data interconnects in

the 1550 nm band", *IEEE Photonics Technology Letters*, vol. 11 (5), pp. 584-586, 2000.

[34] S. Greek, R. Gupta, K. Hjort, "Mechanical considerations in the design of a micromechanical", *Journal of Microelectromechanical Systems*, vol. 8 (3), pp. 328-334, 1997.

[35] J. Pfeiffer, R. Riemenschneider, K. Mutamba, J. Peerlings, B.E. Afshar, H. Baaser, D. Grunert, P. Meissner, H.L. Hartnagel, "Indium Phosphide based bulk micromachining technology for micromechanically tunable WDM filter applications", *Proceedings of Transducers '99*, Sendai, Japan, pp. 380-383, 1999.

[36] J. H. Correia, M. Bartek, R. F. Wolffenbuttel, "Bulk-micromachined tunable Fabry-Perot microinterferometer for the visible spectral range", *Sensors and Actuators A*, vol. 76 (1-3), pp. 191-196, 1999.

[37] X. M. Zhang, F. Chollet, F.S. Chau, C. Quan, A. Q. Liu, "Polysilicon micromachining 3D mirror integrated frequency tuner for WDM applications", *MOEMS and Miniaturzied Systems, Proceedings of SPIE*, vol. 4178, pp. 278-287, 2000.

[38] P. Viktorovitch, J. L. Leclercq, E. Goutain, D. Rondi, "Design and fabrication of optical microcavities using III-V semiconductor-based MOEMS", *MOEMS and Miniaturzied Systems, Proceedings of SPIE*, vol. 4178, pp. 298-309, 2000.

[39] M. Blomberg, O. Rusanen, K. Keranen, A. Letho, "A silicon microsystem - miniaturized infrared spectrometer", *Proceedings of Transducers '97*, Chicago, USA, pp. 1257-1258, 1997.

[40] J. H. Correia, M. Bartek, R. F. Wolffenbuttel, "High-selectivity single-chip spectrometer in silicon for operation at visible part of the spectrum", *IEEE Transactions on Electron Devices*, vol. 47 (3), pp. 553-559, 2000.

[41] K. Knop, "Optical microspectrometers: How small can they be?", *EOS Topical Meeting Digest Series*, vol. 16, pp. 8-9, 1998.

[42] M. Moser, M. Brunner, H. P. Gauggel, K. H. Gulden, R. Hövel, H. P. Zappe, "VSCELs for Datacom and Sensing Applications", *CSEM Scientific and Technical Report*, CSEM Neuchâtel, Switzerland, p. 28, 1999.

[43] M.-H. Kiang, J. T. Nee, K. Y. Lau, R. S. Muller, "Surface-micromachined diffraction gratings for scanning spectroscopic applications," *Proceedings of Transducers '97*, Chicago, USA, pp. 343-345, 1997.

[44] M. B. Sinclair, M. A. Butler, A. J. Ricco, S. D. Senturia, "Synthetic spectra: a tool for correlation spectroscopy", *Applied Optics*, vol. 36 (15), pp. 3342-3348, 1997.

[45] E. S. Hung, S. D. Senturia, "Extending the travel range of analog-tuned electrostatic actuators", *Journal of Microelectromechanical Systems*, vol. 8 (4), pp. 497-505, 1999.

[46] G. B. Hocker, D. Younger, E. Deutsch, A. Volpicelli, S. Senturia, M. Butler, M. Sinclair, T. Plowman, A. J. Ricco, "The Polychromator: A programmable MEMS diffraction grating for synthetic spectra," *Solid-State Sensor and Actuator Workshop*, Hilton Head Island South Carolina, pp. 89-91, 2000.

[47] C. Müller, J. Mohr, "Micropectrometer fabricated by the LIGA process", *Interdisciplinary Science Reviews*, vol. 18 (3), pp. 273-279, 1993.

[48] W. Bacher, W. Menz, J. Mohr, C. Müller, W.K. Schomburg, "Die LIGA-Technik und ihr Potential für Mikrosysteme", *Naturwissenschaften*, vol. 81 (12), pp. 536-545, 1994.

[49] W. Bacher, W. Menz, J. Mohr, "The LIGA technique and its potential for microsystems - a survey", *IEEE Transactions on Industrial Electronics*, vol. 42 (5), pp. 431-441, 1995.

[50] H. Teichmann, M. T. Gale, J. M. Raynor, P. Sietz, H. Schütz, R. Stutz, J. Pedersen, S. Westenhöfer, D. Hiller, "Miniature replicated spectrometers for the visible", *EOS Topical Meeting Digest Series*, vol. 16, pp. 50-51, 1998.

[51] G. M. Yee, N. I. Maluf, G. T. A. Kovacs, "A low-cost micromachined, integrated CMOS spectrometer", *Proceedings of Transducers '99*, Sendai, Japan, pp. 1882-1883, 1999.

[52] T. A. Kwa, R. F. Wolffenbuttel, "Integrated grating / detector array fabricated in silicon using micromachining techniques", *Sensors and Actuators A*, vol. 31, pp. 259-266, 1992.

[53] H. Hogan, "Miniature microscope clears things up", *Biophotonics International*, vol. 7, pp. 20, 2000.

[54] D. L. Dickensheets and G. S. Kino, "Silicon-Micromachined scanning confocal optical microscope", *J. Microelectromech. Systems*, vol. 7, pp. 38-47, 1998.

[55] R. Buser, N. F. de Rooij, H. Tischauser, A. Dommann, and G. Staufert, "Biaxial scanning mirror activated by bimorph structures for medical applications", *Sensors & Actuators A*, vol. 31, pp. 29-34, 1992.

[56] E. Baumann, "The Fischer large screen projection system", *Journal of the SMPTE*, vol. 60, pp. 351-356, 1953.

[57] J. A. van Raalte, "A new Schlieren light valve for television projection", *Applied Optics*, vol. 9, pp. 2225-2230, 1970.

[58] R. N. Thomas, J. Guldenberg, H. C. Nathanson, and P. R. Malmberg, "The Mirror-matrix tube: a novel light valve for projection displays", *IEEE Transactions on electron devices*, vol. ED-22, pp. 765-775, 1975.

[59] J. Guldberg, H. C. Nathanson, D. L. Balthis, and A. S. Jensen, "An aluminium/SiO_2/silicon-on-sapphire light valve matrix or projection displays", *Applied Physics Letters*, vol. 26, pp. 391-393, 1975.

[60] K. E. Petersen, "Silicon torsional scanning mirror", *IBM J. Res. Develop.*, vol. 24, pp. 631-637, 1980.

[61] K. E. Petersen, "Silicon as a mechanical material", *Proceedings of the IEEE*, vol. 77, pp. 420-457, 1982.

[62] J. Kränert, D. Deter, T. Gessner, and W. Dötzel, "Laser Display Technology", *Proceedings of MEMS '98*, Heidelberg, Germany, 1998, pp. 99-104.

[63] W. Lang, H. Pavlicek, Th. Marx, H. Scheithauer, and B. Schmidt, "Electrostatically actuated micromirrors devices in silicon technology", *Sensors & Actuators*, vol. 74, pp. 216-218, 1999.

[64] H. Toshiyoshi and H. Fujita, "An electrostatically operated torsion mirror for optical switching device", presented at Transducers'95, Eurosensors IX, Stockholm, Sweden, 1995, pp. 297-300.

[65] T. Usuda, "Operational characteristics of electrostatically driven torsional resonator with two degrees of freedom", *Sensors & Actuators A*, vol. 64, pp. 255-257, 1998.

[66] T. Usuda and A. Umeda, "An electrostatically driven torsional resonator with two degrees of freedom", *Int. J. Japan Soc. Prec. Eng.*, vol. 31, pp. 135-140, 1997.

[67] H. Schenk, P. Dürr, and H. Kück, "Micro Mechanical Scanning Mirror", Fraunhofer Institut für Mikroelektronische Schaltungen und Systeme, Anual Report, 1998.

[68] H. Schenk, P. Dürr, and H. Kück, "A novel electrostatically driven torsional actuator", presented at MOEMS, Mainz, D, 1999, pp. 3-10.

[69] G. A. Feather, "Digital micromirror device for projection display", *SPIE*, vol. 2407, pp. 90-95, 1995.

[70] L. J. Hornbeck, "Current status of the Digital micromirror device (DMD) for projection television applications", presented at International Electron Devices Meeting 1993, Washington, DC, USA, 1993.

[71] G. T. A. Kovacs, *Micromachined transducers sourcebook*: WCB McGraw-Hill, 1998.

[72] B. Wagner, K. Reimer, A. Maciossek, and U. Hofmann, "Infrared micromirror array with large pixel size and large deflection angle", presented at Transducers '97, Chicago, USA, pp. 75-78, 1997.

[73] J. Bühler, F.-P. Steiner, R. Hauert, and H. Baltes, "Linear array of complementary metal oxide semiconductor double-pass metal micromirrors", *Optical Engineering*, vol. 36, pp. 1391-1398, 1997.

[74] S.-W. Chung and Y.-K. Kim, "Measurements of a fabricated micro mirror using a lateral effect position-sensitive photodiode", *IEEE Transactions on industrial electronics*, vol. 45, pp. 861-865, 1998.

[75] S.-W. Chung, J.-W. Shin, Y.-K. Kim, and B.-S. Han, "Design and fabrication of micro mirror supported by electroplated nickel posts", *Proceedings of Transducers '95, Eurosensors IX*, Stockholm, Sweden, 1995, pp. 312-315.

[76] V. P. Jaecklin, C. Linder, N. F. de Rooij, J.-M. Moret, and R. Vuilleumier, "Line-adressable torsional micromirrors for light modulator arrays", *Sensors & Actuators A*, vol. 41-42, pp. 324-329, 1994.

[77] M. Fischer, H. Graef, and W. von Münch, "Electrostatically deflectable polysilicon torsional mirrors", *Sensors & Actuators A*, vol. 44, pp. 83-89, 1994.

[78] M. Fischer, M. Nägele, D. Eichner, C. Schölhorn, and R. Strobel, "Integration of surface micromachined polysilicon mirrors and a standard CMOS process", *Proceedings of Transducers '95, Eurosensors IX*, Stockholm, Sweden, 1995, pp. 305-307.

[79] O. Solgaard, F. S. Andejas, and D. M. Bloom, "Deformable grating optical modulator", *Optics Letters*, vol. 17, pp. 688-690, 1992.

[80] R. W. Cohn, "Real-time multispot beam steering with electrically controlled spatial light modulators", *SPIE*, vol. 3131, pp. 145-155, 1997.

[81] A. A. Yasseen, S. W. Smith, F. L. Merat, and M. Mehregany, "Diffraction grating scanners using polysilicon micromotors", *IEEE Journal of selected Topics in Quantum Electronics*, vol. 5, pp. 75-82, 1999.

[82] Ch. Jongeneel, "From mirror to projector", DIMES of the Electrical Engineering Faculty of Delft University, Delft, The Netherlands, Outlook, 1996.

[83] T. G. Bifano, R. K. Mali, J. K. Dorton, J. Perreault, N. Vandelli, M. N. Horenstein, and D. A. Castanon, "Continuous-membrane surface-micromachined silicon deformable mirror", *Optical Engineering*, vol. 36, pp. 1354-1360, 1997.

[84] R. S. Muller and K. Y. Lau, "Surface-Micromachined Microoptical Elements and Systems", *Proceedings of the IEEE*, vol. 86, pp. 1705-1720, 1998.

[85] M. C. Wu, "Micromachining for optical and optoelectronic systems", *Proceedings of the IEEE*, vol. 85, pp. 1833-1856, 1997.

[86] Sandia National Laboratories, "Microsystems S&T and Components", *http://www.mems.sandia.gov/*, 2000.

[87] V. M. Bright, J. H. Comtois, D. E. Sene, J. R. Reid, S. C. Gustafson, and E. A. Watson, "Realizing micro-opto-electro-mechanical devices through a commercial surface-micromachining", *SPIE*, vol. 2687, pp. 34-46, 1996.

[88] K. S. J. Pister, M. W. Judy, S. R. Burgett, and R. S. Fearing, "Microfabricated hinges", *Sensors & Actuators A*, vol. 33, pp. 249-256, 1992.

[89] M. J. Daneman, N. C. Tien, O. Solgaard, K. Y. Lau, and R. S. Muller, "Actuated micromachined microreflector with two degrees of freedom for integrated optical systems", presented at Integrated Photonics Research, Boston, 1996, pp. 541-544.

[90] E. J. Garcia and J. J. Sniegowski, "Surface micromachined microengine", *Sensors & Actuators A*, vol. 48, pp. 203-214, 1995.

[91] P. M. Hagelin and O. Solgaard, "Optical Raster-Scanning Displays Based on Surface-Micromachined Polisilicon Mirrors", *IEEE Journal of selected Topics in Quantum Electronics*, vol. 5, pp. 67-74, 1999.

[92] A. Schroth, C. Lee, S. Matsumoto, M. Tanaka, and R. Maeda, "Application of sol-gel deposited thin PZT film for actuation of 1D and 2D scanners", *Proceedings of MEMS '98*, Heidelberg, Germany, 1998, pp. 402-407.

[93] Y. Ohtuka, H. Nishikawa, T. Koumura, and T. Hattori, "Two-dimensional Optical Scanner Applying a Torsional Resonator with two Degrees of Freedom", presented at MEMS 95, 1995, pp. 306-309.

[94] Y. Huang, H. Zhang, and E. S. Kim, "Piezoellectrically actuated microcantilever for actuated mirror array application", presented at Solide-State Sensor and Actuator Workshop, Hilton Head, South Carolina, USA, 1996, pp. 191-195.

[95] R. A. Miller and G. W. Burr, "A magnetically actuated MEMS scanning mirror", *SPIE*, vol. 2687, pp. 47-52, 1996.

[96] R. A. Miller, Y.-C. Tai, G. Xu, J. Bartha, and F. Lin, "An electromagnetic MEMS 2 x 2 fiber optic bypass switch", *Proceedings of Transducers '97*, Chicago, USA, pp. 89-92, 1997.

[97] L. O. S. Ferreira and S. Moehlecke, "A silicon micromechanical galvanometric scanner", *Sensors & Actuator*, vol. 73, pp. 252-260, 1999.

[98] H. Miyajima, N. Asaoka, M. Arima, Y. Minamoto, K. Murakami, K. Tokuda, and K. Matsumoto, "An electromagnetic optical scanner with polymide-based hinges", *Proceedings of Transducers '99*, Sendai, Japan, 1999, pp. 372-375.

[99] A. Garnier, T. Bourouina, H. Fujita, T. Hiramoto, E. Orsier, and J.-C. Peuzin, "Nonlinear behavior of a magnetic microactuator", presented at IEEE, Emerging technologies and factory automation, Barcelona, Catalonia, Spain, 1999, pp. 393-396.

[100] T. Bourouina, A. Garnier, H. Fujita, T. Hiramoto, E. Orsier, and J.-C. Peuzin, "Mechanical characterization of magnetostrictively actuated micro-resonators", presented at Design, Characterization and Packaging for MEMS and Microelectronics, Queensland, Australia, 1999, pp. 411-420.

[101] W. D. Cowan and V. M. Bright, "Thermally actuated piston micromirror array", *SPIE*, vol. 3131, pp. 260-271, 1997.

[102] J. Bühler, J. Funk, O. Paul, H. Baltes, F.-P. Steiner, "Thermally actuated CMOS micromirrors", *Sensors and Actuators A*, vol. 45 (1-3), pp. 572-575, 1995.

[103] M. E. Motamedi, S. Park, A. Wang, M. S. Dadkhah, A. P. Andrews, H. O. Marcy, M. Khoshnevisan, A. E. Chiou, "Development of micro-electro-mechanical optical scanner", *Optical Engineering*, vol. 36 (5), pp. 1346-1353, 1997.

[104] Ch. Alépée, "Technologies for high-temperature silicon microreactors", *Micro-engineering Department, Institute of Microsystems,* Swiss Federal Institute of Technology (EPFL), Lausanne, Switzerland, 2000.

[105] G. Lammel, "New micromachining technologies using porous silicon", *Micro-engineering Department, Institute of Microsystems,* Swiss Federal Institute of Technology (EPFL), Lausanne, Switzerland, 2001.

[106] M. E. Motamedi, M. C. Wu, K. S. J. Pister, "Micro-opto-electro-mechanical devices and on-chip processing", *Optical Engineering*, vol. 36 (5), pp. 1282-1297, 1997.

[107] M. Madou, "Fundamentals of Microfabrication", CRC Press, Boca Raton, Florida, USA, ISBN 0-8493-9451-1, 1997.

[108] P. F. van Kessel, L. J. Hornbeck, R. E. Meier, M. R. Douglas, "A MEMS-based projection display", *Proceedings of the IEEE*, vol. 86 (8), pp. 1687-1704, 1998.

[109] J.-W. Shin, S.-W. Chung, Y.-K. Kim, B. K. Choi, "Design and fabrication of micromirror array supported by vertical springs", *Proceedings of Transducers '97*, Chicago, pp. 201-204, 1997.

[110] B. Wagner, K. Reimer, A. Maciossek, U. Hofmann, "Infrared micromirror array with large pixel size and large deflection angle", *Proceedings of Transducers '97*, Chicago, pp. 75-78, 1997.

[111] G. Perregaux, P. Weiss, B. Kloeck, H. Viulliomenet, J.-P. Thiébaud, "High-speed micro-electromechanical light modulation array", *Proceedings of Transducers '97*, Chicago, pp. 71-74, 1997.

[112] M.-H. Kiang, O. Solgaard, R. S. Muller, and K. Y. Lau, "Micromachined polysilicon microscanners for barcode readers," *IEEE Photon. Technology Letters*, vol. 8 (12), pp. 1707-1709, 1996.

[113] M.-H. Kiang, O. Solgaard, K. Y. Lau, R. S. Muller, "Polysilicon optical microscanners for laser scanning displays", *Sensors and Actuators A,* vol. 70, pp. 195-199, 1998.

[114] S. Schweizer, S. Calmes, M. Laudon and Ph. Renaud, "Thermally actuated optical microscanner with large angle and low consumption", *Sensors & Actuators A*, vol. 76, pp. 470-477, 1999.234

[115] S. Schweizer, P. Cousseau, S. Schiesser, G. Lammel, S. Calmes, and Ph. Renaud, "Thermally actuated microprojector for optical display applications", *MOEMS and Miniaturzied Systems, Proceedings of SPIE*, vol. 4178, pp. 165-175, 2000.

[116] S. Schweizer, P. Cousseau, G. Lammel, S. Calmes, and Ph. Renaud, "Two-dimensional thermally actuated optical microprojector", *Sensors & Actuators A*, vol. 85, pp. 424-429, 2000.

[117] H. Urey, D. W. Wine, T. D. Osborn, "Optical performance requirements for MEMS-scanner based microdisplays", *MOEMS and Miniaturzied Systems, Proceedings of SPIE*, vol. 4178, pp. 176-185, 2000.

[118] T. Gessner, S. Kurth, C. Kaufmann, J. Markert, A. Ehrlich, W. Dötzel, "Micromirrors and micromirror arrays for scanning applications", *MOEMS and Miniaturzied Systems, Proceedings of SPIE*, vol. 4178, pp. 338-347, 2000.

[119] H. Schenk, P. Dürr, D. Kunze, H. Lakner, H. Kück, "Design and modelling of large deflection micromechanical 1D and 2D scanning mirrors", *MOEMS and Miniaturzied Systems, Proceedings of SPIE*, vol. 4178, pp. 116-124, 2000.

[120] D. W. Wine, M. P. Heisel, L. Jenkins, H. Urey, T. D. Osborn, "Perfomance of a biaxial MEMS-based scanner for microdisplay applications", *MOEMS and Miniaturzied Systems, Proceedings of SPIE*, vol. 4178, pp. 186-196, 2000.

[121] M. Ohring, *The materials science of thin films,* Academic Press Inc., NY, 1992.

[122] S. Wolf and R. N. Tauber, *Silicon processing for VLSI Era*, vol. 1, Lattice Press, Sunset Beach, CA, 1987.

[123] D. W. Hoffman and J. A. Thornton, "Internal stresses in Cr, Mo, Ta, and Pt films deposited by sputtering from a planar magnetron source", *J. Vac. Sci. Technol.*, vol. 20, pp. 355-358, 1982.

[124] A.-J. Y. Villarceau, "Recherches sur le mouvement et la compensation des chronometres", Annales de l'Observatoire imperial de Paris, 1863.

[125] R. F. Bunshah, *Handbook of deposition technologies for films and coatings*, 2 ed., New Jersey, USA: Noyes Publications, Park Ridge, 1994.

[126] P. Delobelle, "Rapport sur les essais de nanoindentation realise sur la structure bimorphe film-mince de l'EPFL", LMA, CNRS, Université de Franche-Comte, Besançon, Franche-Comte, France, 1998.

[127] K. Gieck and R. Gieck, *Formulaire technique*, Germering, Deutschland: Gieck Verlag, 1990.

[128] D. R. Lide, *CRC Handbook of chemistry and physics*, 74th ed., CRC Press, Inc., Boca Raton, USA, 1993-1994.

[129] W. Riethmüller and W. Benecke, "Thermally excited silicon microactuators", *IEEE Transactions on electron devices*, vol. 35, pp. 758-763, 1988.

[130] F. Moret and M. H. Geiser, "Determination of the radius of micro-mirror", IRO (Institut de Recherche en Opthalmologie), Ecole d'Ingénieurs du Valais, Sion, Valais, Switzerland, 2000.

[131] F. Moret and M. H. Geiser, "Radius and geometrical properties of scanning mirror obtained by reflection pattern analysis", IRO (Institut de Recherche en Opthalmologie), Ecole d'Ingénieurs du Valais, Sion, Valais, Switzerland, 2000.

[132] W.- H. Chu, M. Mehregany, and R. L. Mullen, "Analysis of tip deflection and force of a bimetallic cantilever microactuator", *J. Microelectromech. Microeng.*, vol. 3, pp. 4-7, 1993.

[133] S. Timoshenko, "Analysis of Bi-metal Thermostats", *Journal of Optical Society of America*, vol. 11, pp. 233-255, 1925.

[134] G. Greitmann, "Micromechanical tactile gripper system for micro assembly", *Institute of Mechanics,* Swiss Federal Institute of Technology Zürich (ETH Zürich), Zurich, Switzerland,1998.

[135] WebElements Ltd, "WebElements Periodic Table", *http://www.webelements.com/*, 2000.

[136] W. Fang and C.-Y. Lo, "On the thermal expansion coefficients of thin films", *Sensors & Actuators A*, vol. 84, pp. 310-14, 2000.

[137] W. Fang, H.-C. Tsai, and C.-Y. Lo, "Determining thermal expansion coefficients of thin films using micromachined cantilevers", *Sensors & Actuators A*, vol. 77, pp. 21-27, 1999.

[138] L. T. Canham (Editor), "Properties of porous silicon", IEE, London, 1997.

[139] W. Lang, A. Drost, P. Steiner, H. Sandmaier, "The thermal conductivity of porous silicon", *Materials Research Society Symposium Proceedings*, vol. 358, pp. 561-566, 1995.

[140] G. Benedetto, L. Boarino, R. Spagnolo, "Evaluation of thermal conductivity of porous silicon layers by a photoacoustic method", *Applied Physics A*, vol. 64, pp. 155-159, 1997.

[141] V. Lysenko, Ph. Roussel, B. Remaki, G. Delhomme, A. Dittmar, D. Barbier, "Study of nano-porous silicon with low thermal conductivity as thermal insulating material", *Journal of Porous Materials*, vol. 7, pp. 177-182, 2000.

[142] D. Schneider, M. D. Trucker, "Non-destructive characterization and evaluation of thin films by laser-induces ultrasonic surface waves", *Thin Solid Films*, vol. 290-291, pp. 305-311, 1996.

[143] L. Kiesewetter, J.-M. Zhang, D. Houdeau, A. Steckenborn, "Determination of Young's moduli of micromechanical thin films using the resonance method", *Sensors and Actuators A*, vol. 35, pp. 153-159, 1992.

[144] Reference overview: http://mems.isi.edu/mems/materials/

[145] H. Kuchling, "Physik", 18th edition, VEB Fachbuchverlag, Leipzig, Germany, 1987.

[146] L. L. Chu, L. Que, Y. B. Gianchandani, "Temperature coefficients of material properties for electrodeposited MEMS", *Proceedings of the 14th IEEE International Conference on MEMS*, Interlaken, Switzerland, 2001.

[147] R. J. M. da Fonseca, J. M. Saurel, A. Foucaran, J. Camassel, E. Massone, T. Taliercio, Y. Boumaiza, "Acoustic investigation of porous silicon layers", *Journal of Materials Science*, vol. 30, pp. 35-39, 1995.

[148] H. Czichos, *Die Grundlagen der Ingenieurwissenschaften*, Berlin, Germany: Springer Verlag - Redaktion Hütte, 1996.

[149] F. P. Incropera, D. P. DeWitt, "Fundamentals of heat and mass transfer", 4th edition, John Wiley & Sons, New York, 1996.

[150] L.-A. Liew, A. Tuantranont, V. M. Bright, "Modeling of thermal actuation in a bulk-micromachined CMOS micromirror", *Microelectronics Journal*, vol. 31, pp. 791-801, 2000.

[151] S. P. Parker, *Dictionary of scientific and technical terms*, 5th ed, McGraw-Hill, Inc., 1994.

[152] J. Söderkvist, "Similarities between piezoelectric, thermal and other internal means of excitating vibrations", *J. Microelectromech. Microeng.*, vol. 3, pp. 24-31, 1993.

[153] Th. S. L. Lammerink, M. Elwenspoek, and J. H. J. Fluitman, "Thermal actuation of clamped silicon microbeams", *Sensors and Materials*, vol. 3, pp. 217-238, 1992.

[154] I. Rotman, "Micro-scanner sur substrat en silicium", IMS, DMT, Swiss Federal Institute of Technology (EPFL), Lausanne, Switzerland, 1996.

[155] M. D. Weber, "Miniature photoacoustic gas sensors", *Micro-engineering Department, Institute of Microsystems,* Swiss Federal Institute of Technology (EPFL), Lausanne, Switzerland, 1999.

[156] C. H. Mastrangelo and C. H. Hsu, "Mechanical stabilityand adhesion of microstructures under capillary forces. I. Basic theory", *J. Microelectromech. Systems*, vol. 2, pp. 33-43, 1993.

[157] C. H. Mastrangelo and C. H. Hsu, "Mechanical stabilityand adhesion of microstructures under capillary forces. II. Exeriments", *J. Microelectromech. Systems*, vol. 2, pp. 44-55, 1993.

[158] N. Takeshima, K. J. Gabriel, M. Ozaki, J. Takahashi, H. Horigushi, and H. Fujita, "Electrostatic parallelogram actuators", *Proceedings of Transducers '91*, pp. 63-66, 1991.

[159] T. Inoue and H. Osatake, "A new drying method of biological specimens for SEM: the t-buthyl alcohol freeze-drying method", *Arch. Histol. Cytol.*, vol. 51, pp. 53-59, 1988.

[160] D. Landolt, "Electrochemical Engineering Aspects of Microfabrication", *Proceedings of the second International symposium on Electrochemical Microfabrication*, vol. 94-32, pp. 141-153, 1994.

[161] H.-Y. Lin and W. Fang, "Rib-reinforced micromachined beam and its applications", *J. Microelectromech. Microeng.*, vol. 10, pp. 93-99, 2000.

[162] S. Schweizer, P. Cousseau, G. Lammel, S. Calmes, and Ph. Renaud, "Two-dimensional thermally actuated optical microscanner", *Proceedings of Eurosensors XIII*, The Hague, NL, pp. 11-12, 1999.

[163] D. C. O'Shea, *Elements of modern optical design*. New York, USA: John Wiley & Sons, Inc., 1985.

[164] E. Hecht, *Optics*, 3rd ed., Addison Wesley Longman, Inc., 1998.

[165] A. Uhlir, jr., "Electrolytic Shaping of Germanium and Silicon", *The Bell systems technical journal*, vol. 35, pp. 333-347, 1956.

[166] D. R. Turner, "Electropolishing Silicon in Hydrofluoric Acid Solutions", *Journal of the Electrochemical Society*, vol. 105, pp. 402-408, 1958.

[167] D. R. Turner, "The Electrochemistry of Semiconductors", p. 176, Editor: P. J. Holmes, Academic Press, London, 1962.

[168] R. L. Meek, "Anodic dissolution of N+ Silicon", *Journal of the Electrochemical Society*, vol. 118, pp. 437-442, 1971.

[169] M. J. J. Theunissen, "Etch channel formation during anodic dissolution of n-type silicon an aqueous hydrofluoric acid", *Journal of the Electrochemical Society*, vol. 119, pp. 351-360, 1972.

[170] L. T. Canham, "Silicon quantum wire array fabrication by electrochemical and chemical dissolution of wafers", *Applied Physics Letters*, vol. 57 (10), pp. 1046-1048, 1990.

[171] V. Parkhutik, "Analysis of publications on porous silicon: from photoluminescence to biology", *Journal of Porous Materials*, vol. 7, pp. 363-366, 2000.

[172] A. G. Cullis, L. T. Canham and P. D. J. Calcott, "The structural and luminescence properties of porous silicon", *Journal of Applied Physics*, vol. 82, pp. 909-965, 1997.

[173] J.-C. Vial and J. Derrien (Editors) "Porous silicon in science and technology", *Winter School Les Houches*, Springer-Verlag, Berlin, Les Editions de Physique, Les Ulis, France, 1994.

[174] G. Amato (Editor), "Structural and optical properties of porous silicon nanostructures", Gordon & Breach, New York, 1997.

[175] V. Lehmann, W. Hönlein, H. Reisinger, A. Spitzer, H. Wendt and J. Willer, "A novel capacitor technology based on porous silicon", *Thin Solid Films*, vol. 276, pp. 138-142, 1997.

[176] K. Sakaguchi *et al.*, "Extremely high selective etching of porous Si for single etch-stop bonding-and-etch-back silicon-on-insulator", *Japanese Journal of Applied Physics*, vol. 34, pp. 842-847, 1995.

[177] H. Münder, St. Frohnhoff, M. G. Berger, M. Marso, M. Thönissen, R. Arens-Fischer, H. Lüth, "Extended quantum model for porous silicon formation", *Materials Research Society Symposium Proceedings*, vol. 358, pp. 315-320, 1995.

[178] P. Schmuki, L. E. Erickson, D. J. Lockwood, J. W. Fraser, G. Champion, H. J. Labbé, "Formation of light emitting porous GaAs micropatterns", *Applied Physics Letters*, vol. 72, no. 9, pp. 1039-1041, 1998.

[179] O. Jessensky, F. Müller, U. Gösele, "Microstructure and luminescence of electrochemically etched porous SiC", *Thin Solid Films*, vol. 297, pp. 224-228, 1997.

[180] G. Wilkes, "Hydrofluoric Acid Burns", emedicine, 2000, http://www.emedicine.com/emerg/topic804.htm

[181] S. Hilbrich, W. Theiss, R. Arens-Fischer, O. Glück, M. G. Berger, "The influence of the doping level on the optical properties of porous silicon", *Thin Solid Films*, vol. 276, pp. 231-234, 1996.

[182] V. Lehmann, "The physics of macroporous silicon formation", *Thin Solid Films*, vol. 255, pp. 1-4, 1995.

[183] G. Lérondel, R. Romestain, S. Barret, "Roughness of the porous silicon dissolution interface", *Journal of Applied Physics*, vol. 81, no. 9, pp. 6171-6178, 1997.

[184] M. Thönissen, M. G. Berger, R. Arens-Fischer, O. Glück, M. Krüger, H. Lüth, "Illumination-assisted formation of porous silicon", *Thin Solid Films*, vol. 276, pp. 21-24, 1996.

[185] V. V. Doan, M. J. Sailor, "Photolithographic fabrication of micron-dimension porous Si structures exhibiting visible luminescence", *Applied Physics Letters*, vol. 60, no. 5, pp. 619-620, 1991.

[186] G. Lerondel, R. Romestain, J. C. Vial, M. Thonissen, "Porous silicon lateral superlattices", *Applied Physics Letters*, vol. 71, no. 2, pp. 196-198, 1997.

[187] S. Setzu, P. Ferrand, R. Romanstain, "Optical properties of multilayered porous silicon", *Materials Research and Engineering*, vol. B 69-70, pp. 34-42, 2000.

[188] R. Arens-Fischer, "Charakterisierung und Optimierung der Materialeigenschaften von porösem Silicium für aktive und passive optische Bauelemente", PhD thesis at the Rheinisch-Westfälische Technische Hochschule (RWTH) Aachen, Germany, 1999.

[189] U. Grüning, A. Yelon, "Capillary and Van der Waals forces and mechanical stability of porous silicon", *Thin Solid Films*, vol. 255, pp. 135-138, 1995.

[190] O. Belmont, C. Faivre, D. Bellet, Y. Brechet, "About the origin and the mechanism involved in the cracking of highly porous silicon layers under capillary stress", *Thin Solid Films*, vol. 276, pp. 219-222, 1996.

[191] J. v. Behren, L. Tsybeskov, P. M. Fauchet, "Preparation and characterization of ultrathin porous silicon films", *Applied Physics Letters*, vol. 66, no. 13, pp. 1662-1664, 1995.

[192] K. Rauscher, J. Voigt, I. Wilke, K. Wilke, "Chemische Tabellen und Rechentafeln für die analytische Praxis", 8th edition, VEB Deutscher Verlag für Grundstoffindustrie, Leipzig, 1986.

[193] W. Lang, P. Steiner, A. Richter, K. Marusczyk, G. Weimann, H. Sandmaier, "Application of porous silicon as a sacrificial layer", *Sensors and Actuators A*, vol. 43, pp. 239-242, 1994.

[194] W. Lang, P. Steiner, H. Sandmaier, "Porous silicon: a novel material for microsystems", *Sensors and Actuators A*, vol. 51, pp. 31-36, 1995.

[195] F. Hedrich, S. Billat, W. Lang, "Structuring of membrane sensors using sacrificial porous silicon", *Sensors and Actuators A*, vol. 84, pp. 315-323, 2000.

[196] C.-S. Lee, J.-D. Lee, C.-H. Han, "A new wide-dimensional freestanding microstructure fabrication technology using laterally formed porous silicon as a sacrificial layer", *Sensors and Actuators A*, vol. 84, pp. 181-185, 2000.

[197] T. E. Bell, P. T. J. Gennissen, D. DeMunter, M. Kuhl, "Porous silicon as sacrificial material", *Journal of Micromechanics and Microengineering*, vol. 6, pp. 361-369, 1996.

[198] A. Zeitschel, A. Friedberger, W. Welser, G. Müller, "Breaking the isotropy of porous silicon formation by means of current focusing", *Sensors and Actuators A*, vol. 74, pp. 113-117, 1999.

[199] M. G. Berger , S. Frohnhoff, R. Arens-Fischer, M. Thönissen, C. Dieker, H. Münder, H. Lüth, "Fabrication and properties of multilayer porous silicon filters", *Proceedings of SPIE*, Volume 2253: Optical Interference Coatings, pp. 865-871, 1994.

[200] St. Frohnhoff, M. G. Berger, M. Thönissen, C. Dieker, L. Vescan, H. Münder, H. Lüth, "Formation techniques for porous Si superlattices", *Thin Solid Films*, vol. 255, pp. 59-62, 1995.

[201] M. G. Berger, M. Thönissen, R. Arens-Fischer, H. Münder, H. Lüth, M. Arntzen, W. Theiss, "Investigation and design of optical properties of porosity superlattices", *Thin Solid Films*, vol. 255, pp. 313-316, 1995.

[202] M. Thönissen, M. G. Berger, M. Krüger, S. Billat, R. Arens-Fischer, O. Glück, H. Lüth, S. Hilbrich, W. Theiss, P. Grosse, "Optical interference filters made of porous silicon", *Materials Research Society Symposium Proceedings*, vol. 431, pp. 373-378, 1996.

[203] M. G. Berger, H. Münder, S. Frohnhoff, H. Lüth, M. Thönissen, "Optoelektronisches und optisches Bauelement", patent, DE: P 43 19 413.3-33 (14.06.93) and PCT: DE94/00635 (08.06.94) (EP, JP, US).

[204] G. Vincent, "Optical properties of porous silicon superlattices", *Applied Physics Letters*, vol. 64, no. 18, pp. 2367-2369, 1994.

[205] L. Pavesi, "Porous silicon dielectric multilayers and microcavities", Rivista del Nuovo Cimento, vol. 20 (10), pp. 1-76, 1997.

[206] A. Thelen, "Design of optical interference coatings", McGraw-Hill, New York, 1st edition, 1989.

[207] A. R. Offer, J. Bland-Hawthorn, "Rugate filters for OH-suppressed imaging at near-infrared wavelengths", *Monthly Notices of the Royal Astronomical Society*, vol. 299 (1), pp. 176-188, 1998.

[208] P. L. Swart, A. P. Kotzé, B. M. Lacquet, "Effects of the nature of the starting population on the properties of rugate filters designed with the genetic algorithm", *Joural of Lightwave Technology*, vol. 18 (6), pp. 853-859, 2000.

[209] W. H. Southwell, R. L. Hall, "Rugate filter sidelobe suppression using quintic and rugated matching layers", *Applied Optics*, vol. 28, no. 14, pp. 2949-2951, 1989.

[210] W. H. Southewell, "Using apodization functions to reduce sidelobes in rugate filters", *Applied Optics*, vol. 28, no. 23, pp. 5091-5094, 1989.

[211] W. Theiss, S. Henkel, M. Arntzen, "Connecting microscopic and macroscopic properties of porous media: choosing appropriate effective medium concepts", *Thin Solid Films*, vol. 255, pp. 177-180, 1995.

[212] W. Theiss, "The dielectric function of porous silicon - how to obtain it and how to use it", *Thin Solid Films*, vol. 276, pp. 7-12, 1996.

[213] W. Theiss, "Optical properties of porous silicon", *Surface Science Reports*, vol. 29, pp. 91-192, 1997.

[214] H. Münder, M. G. Berger, H. Lüth, U. Rossow, U. Frotscher, W. Richter, R. Herino, M. Ligeon, "The influence of nanocrystals on the dielectric function of porous silicon", *Applied Surface Science*, vol. 63, pp. 57-61, 1993.

[215] U. Rossow, U. Frotscher, C. Pietryga, W. Richter, D. E. Aspnes, "Interpretation of the dielectric function of porous silicon layers", *Applied Surface Science*, vol. 102, pp. 413-416, 1996.

[216] W. Theiss, "The use of effective medium theory in optical spectroscopy", *Festkörperprobleme*, vol. 33, pp. 149-176, 1994.

[217] E. D. Palik, "Handbook of optical constants of solids", Academic Press, Orlando, 1985.

[218] M. Thönissen, S. Billat, M. Krüger, H. Lüth, M. G. Berger, U. Frotscher, U. Rossow, "Depth inhomogeneity of porous silicon layers", *Journal of Applied Physics*, vol. 80 (5), pp. 2990-2993, 1996.

[219] M. Thönissen, M. G. Berger, S. Billat, R. Arens-Fischer, M. Krüger, H. Lüth, W. Theiss, S. Hillbrich, P. Grosse, G. Lerondel, U. Frotscher, "Analysis of the depth inhomogeneity of p-PS by reflectance measurements", *Thin Solid Films*, vol. 297, pp. 92-96, 1997.

[220] M. Thönissen, M. G. Berger, W. Theiss, S. Hillbrich, M. Krüger, R. Arens-Fischer, S. Billat, G. Lerondel, H. Lüth, "Depth gradients in porous silicon: How to measure them and how to avoid them", *Materials Research Socienty Symposium*, vol. 452, pp. 431-436, 1997.

[221] S. Billat, M. Thönissen, R. Arens-Fischer, M. G. Berger, M. Krüger, H. Lüth, "Influence of etch stops on the mircostructure of porous silicon layers", *Thin Solid Films*, vol. 297, pp. 22-25, 1997.

[222] M. G. Berger, "Poröses Silizium für die Mikrooptik: Herstellung, Mikrostruktur und optische Eigenschaften von Einzelschichten und Schichtsystemen", PhD thesis at the Rheinisch-Westfälische Technische Hochschule (RWTH) Aachen, Forschungszentrum Jülich, Institut für Schicht- und Ionentechnik, Jül-3219, 1996.

[223] K. Peters, M. Studer, J. Botsis, A. Iocco, H. Limberger, R. Salathé, "Embedded optical fiber Bragg grating sensor in a nonuniform strain field: Measurements and simulations", *Eperimental Mechanics*, vol. 41 (1), pp. 19-28, 2001.

[224] M. Morita, T. Ohmi, E. Hasegawa, M. Kawakami, M. Ohwada, "Growth of native oxide on a silicon surface", *Journal of Applied Physics*, vol. 68 (3), pp. 1272-1280, 1990.

[225] U. Frotscher, U. Russow, M. Ebert, C. Pietryga, W. Richter, M. G. Berger, R. Arens-Fischer, H. Münder, "Investigation of different oxidation processes for porous silicon studied by spectroscopic ellipsometry", *Thin Solid Films*, vol. 276, pp. 36-39, 1996.

[226] M. Krüger, S. Hilbrich, M. Thönissen, D. Scheyen, W. Theiss, H. Lüth, "Suppression of ageing effects in porous silicon interference filters", *Optics Communications*, vol. 146, pp. 309-315, 1998.

[227] J. J. Yon, K. Barla, R. Herino, G. Bomchil, "The kinetics and mechanism of oxide layer formation from porous silicon formed on p-Si substrates", *Journal of Applied Physics*, vol. 62 (3), pp. 1042-1048, 1987.

[228] V. Lehmann, "Porous silicon – a new material for MEMS", *Proceedings of the 9th Annual IEEE International Workshop on MEMS*, San Diego, California, USA, pp. 1-6, 1996.

[229] R. W. Tjerkstra, J. G. E. Gardeniers, M. C. Elwenspoek, A. van den Berg, "Electrochemical fabrication of multi walled micro channels", *Proceedings of the Micro Total Analysis Systems '98 Workshop*, Banff, Canada, pp. 133-136, 1998.

[230] M. J. de Boer, R. W. Tjerkstra, J. W. Berenschot, H. V. Jansen, G. J. Burger, J. G. E. Gardeniers, M. C. Elwenspoek, A. van den Berg, "Micromachining of buried micro channels in silicon", *Journal of Microelectromechanical Systems*, vol. 9 (1), pp. 94-102, 2000.

[231] H. Ohji, P. J. Trimp, P. J. French, "Fabrication of free standing structure using single step electrochemical etching in hydrofluoric acid", *Sensors and Actuators A,* vol. 73, pp. 95-100, 1999.

[232] H. Ohji, P. J. French, K. Tsutsumi, "Fabrication of mechanical structures in p-type silicon using electrochemical etching", *Proceedings of the Transducers '99 Conference*, Sendai, Japan, pp. 1086-1089, 1999.

[233] H. Ohji, P. J. French, S. Izuo, K. Tsutsumi, "Initial pits for electrochemical etching in hydrofluoric acid", *Sensors and Actuators A,* vol. 85, pp. 390-394, 2000.

[234] H. Ohji, P. T. J. Gennissen, P. J. French, K. Tsutsumi, "Fabrication of accelerometer using single-step electrochemical etching for micro

structures (SEEMS)", *Proceedings of the 12th Annual International Conference on MEMS*, Orlando, USA, pp. 61-65, 1999.

[235] G. Lammel, Ph. Renaud, "3D flip-up structure of porous silicon with actuator and optical filter for microspectrometer applications", *Proceedings of the 13th IEEE MEMS 2000 Conference*, Miyazaki, Japan, pp. 132-137, 2000.

[236] G. Lammel, Ph. Renaud, "Free-standing, mobile 3D porous silicon microstructures", *Sensors & Actuators A,* vol. 85, pp. 356-360, 2000.

[237] G. Lammel, S. Schweizer, and Ph. Renaud, "Low cost micro-spectrometer," *Proceedings of SPIE's Micromachining and Microfabrication Symposium,* Santa Clara, California, vol. 4178, pp. 288-295, 2000.

[238] G. Lammel, S. Schweizer, Ph. Renaud, "MEMS infrared gas spectrometer based on a porous silicon tunable optical filter", *Proceedings of the 14th IEEE International Conference on MEMS,* Interlaken, Switzerland, pp. 578-581, 2001.

[239] G. Lammel, S. Schweizer, Ph. Renaud, "Microspectrometer based on a tunable optical filter of porous silicon", *Sensors & Actuators A*, vol. 92, pp. 52-59, 2001.

[240] G. Lammel, European patent application 99117852.6, PCT/EP 00/08655, WO 01/19723, "Procédé de fabrication d'une microstructure intégrée suspendue tridimensionelle, micro-structure intégrée notamment obtenue par ce procédé et élément optique intégré réglable", 1999, 2000, 2001.

[241] P. Steiner, W. Lang, "Micromachining applications of porous silicon", *Thin Solid Films*, vol. 255, pp. 52-58, 1995.

[242] J.-N. Chazalviel, M. Etman, F. Ozanam, "A voltammetric study of the anodic dissolution of p-Si in fluoride electrolytes", *Journal of Electro-analytical Chemistry*, vol. 297, pp. 533-540, 1991.

[243] Th. Dittrich, S. Rauscher, V. Yu. Timoshenko, J. Rappich, I. Sieber, H. Flietner, H. J. Lewerenz, "Ultrathin luminescent nanoporous silicon on n-Si: *p*H dependent preparation in aqueous NH_4F solutions", *Applied Physics Letters*, vol. 67 (8), pp. 1134-1136, 1995.

[244] M. P. Stewart, J. M. Buriak, "Photopatterned Hydrosilylation on porous silicon", *Angewandte Chemie, International Edition*, vol. 37 (23), pp. 3257-3260, 1998.

[245] M. R. Linford, C. E. D. Chidsey, "Alkyl monolayers covalently bonded to silicon surfaces", *Journal of the American Chemical Society*, vol. 115, pp. 12631-12632, 1993.

[246] F. Effenberger, G. Götz, B. Bidlingmeier, M. Wezstein, "Photoactivated preparation and patterning of self-assembled monolayers with 1-alkenes and aldehydes on silicon surfaces", *Angewandte Chemie, International Edition*, vol. 37 (18), pp. 2462-2464, 1998.

[247] P. W. Hoffmann, M. Stelzle, J. F. Rabolt, "Vapor phase self-assembly of fluorinated monolayers on silicon and germanium oxide", *Langmuir*, vol. 13, pp. 1877-1880, 1997.

[248] P. Merkli, "Technologies de masque pour le silicium poreux", semester project summer 1999 at the Institute of Microsystems, Swiss Federal Institute of Technology, Lausanne, (EPFL).

[249] Z. M. Rittersma, "Microsensor applications of porous silicon", PhD thesis at the University of Bremen, Germany, Faculty of Physics and Electrical Engineering, 1999.

[250] A. G. Nassiopoulos, S. Grigoropoulos, L. Canham, A. Halimaoui, I. Berbezier, E. Gogolides, D. Papadimitriou, "Sub-micrometer luminescent porous silicon structures using lithographically patterned substrates", *Thin Solid Films*, vol. 255, pp. 329-333, 1995.

[251] G. Kaltsas, A. G. Nassiopoulos, "Frontside bulk micro-machining using porous silicon technology", *Sensors and Actuators A*, vol. 65, pp. 175-179, 1998.

[252] J. Lichtenberg, G. Lammel, M. Oulevey, E. Verpoorte, Ph. Renaud, N. F. de Rooij, "Fabrication of electrically insulated microchannels in silicion", *Proceedings of the Eurosensors XIV Conference*, Copenhagen, Denmark, pp. 463-466, 2000.

[253] M. Krüger, R. Arens-Fischer, M. Thönissen, H. Münder, M. G. Berger, H. Lüth, S. Hilbrich, W. Theiss, "Formation of porous silicon on patterned substrates", *Thin Solid Films*, vol. 276, pp. 257-260, 1996.

[254] R. de Reus, "Diffusion barriers", Chapter 29 in "Intermetallic compounds", vol. 2, edited by J. Westbrook *et al.*, John Wiley & Sons, pp. 603-635, 1994.

[255] G. Gerlach, W. Dötzel, "Grundlagen der Mikrosystemtechnik", Hanser Verlag, München, Germany, 1st edition, 1997.

[256] S. Schweizer, "Thermally actuated micromirror for one- and two-dimensional optical beam scanning", Ph.D. thesis at the Swiss Federal Institute of Technology Lausanne, Institute for Microsystems, 2001.

[257] J. M. Gere, S. P. Timoshenko, "Mechanics of Materials", 3rd edition, PWS-KENT Publishing Company, Boston, USA, 1990.

[258] Y. H. Ogata, T. Tsuboi, T. Sakka, S. Naito, "Oxidation of porous silicon in dry and wet environments under mild temperature conditions", *Journal of Porous Materials*, vol. 7, pp. 63-66, 2000.

[259] H.-Y. Lin, W. Fang, "Rib-rienforced micromachined beam and its applications", *Journal of Micromechanics and Microengineering*, vol. 10, pp. 93-99, 2000.

[260] J. T. Nee, R. A. Conant, M. R. Hart, R. S. Muller, K. Y. Lau, "Stretched-film micromirrors for improved optical flatness", *Proceedings of the 13th IEEE International Conference on MEMS*, Miyazaki, Japan, pp. 704-709, 2000.

[261] J. W. Suh, Ch. W. Storment, and G. T. A. Kovacs, "Characterization of multi-segment organic thermal actuators", *Proceedings of Transducers '95, Eurosensors IX*, Stockholm, Sweden, pp. 333-336, 1995.

[262] M. L. James, G. M. Smith, J. C. Wolford, and P. W. Whaley, *Vibration of mechanical and structural systems*. New York, USA: Harper & Row, 1989.

[263] M. Del Pedro, "Resistance des materiaux", Notes de cours, EPFL, 1985.

[264] M. Del Pedro and P. Pahud, *Mécanique vibratoire - Système discrets linéaires*. Lausanne, Switzerland, Presses polytechniques et universitaires romandes, 1989.

[265] W. T. Thomson, *Theory of vibration with applications*. London, GB, Sydney, Australia, Wellington, New Zealand: Allen & Unwin, Unwin Hyman Ltd., 1988.

[266] Ch. Gruber, *Mécanique générale*, 1st ed., Lausanne, Switzerland: Presses polytechniques romandes, 1988.

[267] S. Calmes, S. Schweizer, and Ph. Renaud, "A resonating large angle and low consumption micromachined optical scanner", *SPIE*, vol. 3276, pp. 96-102, 1998.

[268] M. Geiser and F. Moret, "Final MINAST report", Ecole d'Ingénieurs du Valais, Sion, Switzerland, 1999.

[269] M. J. Wale and C. Edge, "Self alignement flip-chip assembly of photonic devices with electrical and optical connection", *IEEE Trans. on Components, Hybrids, and Manuf. Tech.*, vol. 13, pp. 780-786, 1990.

[270] H. J. Müller, C. S. Pouillets, "Lehrbuch der Physik", Bd. 2/3 (Optik), Vieweg, Braunschweig, p. 627, 1909.

[271] L. S. Rothmann, "HITRAN Atmospheric radiation measurment", (CD-ROM), 1996, http://www.hitran.com/.

[272] Datasheets of pyroelectric IR-detectors, series LHi, EG&G Heimann Optoelectronics, Wiesbaden, Germany, 2000.

[273] R. Frisch-Fay, *Flexible Bars*, Butterworth and Co. Ltd., 1962.

[274] K. E. Bisshopp and D. C. Drucker, "Large deflection of cantilever beams", *Quartely of Applied Mathematics*, vol. 3, pp. 272-275, 1945.

[275] F. V. Rhode, "Large deflections of cantilever beam with uniformly distributed load", *Quartely of Applied Mathematics*, vol. 11, pp. 337-338, 1953.

[276] J. P. Ryan and M. Steinberg, "WDM and optical Networks: Market Directions", *Optics & Photonics News*, pp. 24-28, 1998.

[277] J. A. Walker, "Telecommunications Applications of MEMS", *mstnews*, pp. 6-9, 2000.

[278] J. A. Walker, "The future of MEMS in telecommunications networks", *J. Microelectromech. Microeng.*, vol. 10, pp. R1-R7, 2000.

[279] R. Allan, "MEMS design gear up for greater commercialization", *Electronic Design*, pp. 85-96, 2000.

[280] L. Y. Lin and E. L. Goldstein, "Light-wave micromachines for optical crossconnects", presented at The 1999 European Conference on Optical Communication, Nice, France, 1999, pp. 114-115.

[281] S. Glöckner, R. Göring, and B. Götz, "Multichannel fiber-optic switches based on MOEM system", *SPIE*, vol. 3226, pp. 76-85, 1997.

[282] P. M. Hagelin, U. Krishnamoorthy, C. M. Arft, J. P. Heritage, and O. Solgaard, "Scalable fiber optic switch using micromachined mirrors", *Proceedings of Transducers '99*, Sendai, Japan, pp. 782-785, 1999.

[283] T. Matsuura, T. Fukami, M. Chabloz, Y. Sakai, S. Izuo, A. Uemura, S. Kaneko, K. Tsutsumi, et al., "Silicon micro optical switching device

with an electromagnetically operated cantilever", *Proceedings of Transducers '99*, Sendai, Japan, pp. 386-389, 1999.

[284] S.-S. Lee, E. Motamedi, and M. C. Wu, "Surface-micromachined free-space fiber optic switches with integrated microactuators for optical fiber communication systems", *Proceedings of Transducers '97*, Chicago, USA, pp. 85-88, 1997.

[285] C. Marxer and N. F. de Rooij, "Reflective modulators and by-pass-switches: two MEMS components for fiber optic communications", *Proceedings of LEOS '98*, Monterey, CA, USA, pp. 27-28, 1998.

[286] N. Anscombe, "European MEMS make a move for the big league", *Opto & Laser Europe*, pp. 26-28, 2000.

SYMBOLS AND ABBREVIATIONS

ABBREVIATIONS

Abbreviation	Definition
AC	Alternating Current
BHF	Buffered HydroFluoric acid
BSAC	Berkeley Sensor & Actuator Center
CCD	Charge Coupled Device
CMI	Center of Microtechnology
CMOS	Complementary Metal Oxide Semiconductor
CRT	Cathode Ray Tube
CVD	Chemical Vapor Deposition
DC	Direct Current
DWDM	Dense Wavelength Division Multiplexed
DMD	Digital Micromirror Display
EIV	Ecole d'Ingénieurs du Valais, Switzerland
EPFL	Ecole Polytechnique Fédérale de Lausanne
FEM	Finite Element Modelling
FWHM	Full Width Half Maximum
GLV	Grating Light Valve
GRIN	GRadient INdex lens
HDTV	High Density TeleVision
HF	Hydrofluoric Acid
HMDS	HexaMethylDiSilazane

Abbreviation	Definition
IC	Integrated Circuit
ICP	Inductively Coupled Plasma
IEEE	Institute of Electrical and Electronics Engineers
IR	InfraRed
IRO	Institut pour la Recherche Ophtalmologique
KOH	potassium hydroxide
LPCVD	Low-Pressure Chemical Vapor Deposition
LCD	Liquid-Crystal Display
LCOS	Liquid-Crystal-On-Silicon
LED	Light Emitting Diode
MEMS	Micro Electro Mechanical System
MINAST	MIcro NAno SysTems
MOEMS	Micro Opto Electro Mechanical System
MTF	Modulation Transfer Function
NEXUS	Network of eXcellence in mUltifunctional microsystemS
NTSC	National Television Standards Committee
PAL	Phase Alternate Lines
PCB	Printed Circuit Board
PSD	Position Sensitive Detector
PVD	Physical Vapor Deposition
QWOT	Quarter Wavelength Optical Thickness Layer
RIE	Reactive Ion Etching
RTA	Rapid Thermal Anneal
QVGA	Quarter Video Graphics Array
SECAM	Sequential Couleur Avec Memoire
SEM	Scanning Electron Microscope

Abbreviation	Definition
SIMS	Secondary Ion Mass Spectroscopy
SIMOX	Separation by Ion iMplantation of OXygen
SOB	Silicon Optical Bench
SOI	Silicon On Insulator
SVGA	Super Video Graphics Array
SXGA	Super Extended Graphics Array
TCR	Temperature Coefficient of Resistance, - of Resistivity
VGA	Video Graphics Array
VRD	Virtual Retinal Display
WDM	Wavelength Division Multiplexing
XGA	eXtended Graphics Array

SYMBOLS

Roman letters

Symbol	Property	Unit
A	surface area	cm^2
a	acceleration *or*	m/s^2
a	mirror shape factor	-
$[B]$	flexibility matrix	-
b	width	m
c	specific heat capacity	J/kg·K
c_0	speed of light in vacuum	m/s
c_{gas}	gas concentration	$1/m^3$
D	mirror diameter, mirror size *or*	m
D	diffusivity constant	m^2/s

Symbol	Property	Unit
d	geometric layer thickness in an optical filter	m
E, E_s	Young's modulus, substrate's Young's modulus	Pa
E'	biaxial Young's modulus, biaxial elastic modulus	Pa
e	unit charge	As
F	force	N
f	frequency	Hz
f_r	raster scanner frequency	Hz
f_f	frame rate	Hz
g	acceleration of gravity	m/s^2
h	thickness	m
I	moment of inertia of a plane area	m^4
I	etch current	A
i	complex number	$\sqrt{-1}$
J	moment of inertia	$kg \cdot m^2$
j	current density	mA/cm^2
K_{sp}	spot size to pixel size ratio	-
K_{os}	overscan constants	-
K_t	beam clipping constant	-
K_{ub}	constant for uni- or bi-directional scanning	-
$[K]$	stiffness matrix	-
k	spring constant, stiffness	kg/s^2
L	length	m
l	cantilever length	m
Δl	length offset	m
M	torque moment	$N \cdot m$
m	mass	kg

Symbol	Property	Unit
N	number of pixels, resolution	-
N	total number of layers of a filter	-
n	number of vertical feed beams	-
n	refractive index	-
P	power	W
P	force, load	N
P	porosity	-
P_{el}	electric power per volume	W/m^3
Q	quality factor	-
Q'	heat transfer rate	W
q'	heat transfer per unit area	W/m^2
r	curvature radius	m
r_{in}, r_{th}	initial and thermal curvature radius	m
R	electrical resistance	Ω
$R(\lambda)$	spectral reflectivity	-
Re	Reynolds number	-
s	thermal sensitivity *or*	° / K
s	vibration sensitivity	s^{-2}
T	temperature	K
T_0, T_{max}	initial temperature, maximal temperature	K
T_{bulk}	temperature of the bulk Si	K
T_{air}	air temperature	K
ΔT	temperature variation	K
$T(\lambda)$	spectral transmissivity	-
V	volume	m^3
t	thickness	m

Symbol	Property	Unit
t_s, t_f	substrate thickness, film thickness	m
U	voltage	V
u	gas velocity	m/s
u	normalized layer thickness	-
V	volume	m^3
v	bending deflection	
v	electrovalence	-
x, y	dimension along wafer surface	m
Z_0	impedance in free space	Ω
z	dimension normal to wafer surface	m

Greek letters

Symbol	Property	Unit
α	thermal expansion coefficient *or*	K^{-1}
α	convection coefficient *or*	$W/m^2 \cdot K$
α	feed beam angle	rad
β	relative temperature coefficient of electrical resistance	K^{-1}
γ	constant for relation current density vs. porosity	-
δ_{max}	maximal mirror deformation	m
ε, $\Delta\varepsilon$	strain, strain variation	-
ε^*, $\Delta\varepsilon^*$, $\Delta\varepsilon_{th}$	residual strain, residual strain difference, thermal strain difference	-
κ	torsion spring constant	N·m
λ	thermal conductivity *or*	W/m·K
λ	wavelength	m
λ_0	central wavelength at normal incidence	m
λ_α	central wavelength at incidence angle α	m

Symbol	Property	Unit
ν	Poisson's ratio	-
ϕ	phase shift of on optical layer	rad
θ, θ_{in}	angle, initial rest angle	rad, °
φ	actuator deflection angle	°
Θ_{max}	maximum mechanical scan amplitude	rad
ρ	mass density	kg/m^3
ρ_{el}	electrical resistivity	μΩ·cm
σ	stress	Pa
σ_{th1}	external thermal stress	Pa
σ_{th2}	thermal stress resulting from the deposition temperature	Pa
σ_e	external stress	Pa
σ_{ext}	extrinsic stress	Pa
σ_i	intrinsic stress	Pa
σ^*	residual stress	Pa
$\sigma(\lambda)$	gas absorption cross section	cm^2
τ	response time	s
υ	kinematic viscosity	m^2/s
ω, ω_0	angular frequency, fundamental angular natural frequency	rad/s
ζ	constant for equation of current density vs. porosity	mA/cm^2

Glossary of terms

A

actuation sensitivity 42
annealing 44, 72
ANSYS 94
apodization function 149
average roughness 67

B

barcode reader 105
bending radius 41
BHF 67
Bragg reflector 148
Bragg's law 153
burn-in 44, 79

C

chromium thin film 78
classification of porous silicon 140
convection 52
cooling 98
crossbar switch 11
current density 142
cut-off frequency 59, 98

D

damping 97
diamond suspension 67
dielectric filter 135
DMD (Digital Mirror Device) 4
doping of Si 140
dry release 70

K

L

M

N

O

P

Q

R

ACKNOWLEDGMENTS

The research reported in this book was supported by the Swiss Federal Institute of Technology Lausanne (EPFL), the Swiss Microtechnology program MINAST and the Center of Microtechnology (CMI) at the EPFL. The authors would like to thank the company Axiome, the EIV (Ecole d'Ing. du Valais), the University of Franche Comte, the group PorSiLux at the Research Center Jülich and all colleagues at the Institute of Microsystems (IMS) at EPFL. Special thanks go to Prof. Steven D. Senturia at the MIT and Mark de Jongh at Kluwer.